T0329741

Advanced Processing and Manufacturing Technologies for Structural and Multifunctional Materials

Advanced Processing and Manufacturing Technologies for Structural and Multifunctional Materials

A Collection of Papers Presented at the 31st International Conference on Advanced Ceramics and Composites
January 21–26, 2007
Daytona Beach, Florida

Editors

Tatsuki Ohji
Mrityunjay Singh

Volume Editors

Jonathan Salem
Dongming Zhu

The American Ceramic Society

BICENTENNIAL
1807
WILEY
2007
BICENTENNIAL

WILEY-INTERSCIENCE
A John Wiley & Sons, Inc., Publication

Published by John Wiley & Sons, Inc., Hoboken, New Jersey.
Published simultaneously in Canada.

For general information on our other products and services or for technical support, please contact our
Customer Care Department within the United States at (800) 762-2974, outside the United States at
(317) 572-3993 or fax (317) 572-4002.

Wiley also publishes its books in a variety of electronic formats. Some content that appears in print may
not be available in electronic format. For information about Wiley products, visit our web site at
www.wiley.com.

Wiley Bicentennial Logo: Richard J. Pacifico

Library of Congress Cataloging-in-Publication Data is available.

ISBN 978-0-470-19638-0

Printed in the United States of America.

10 9 8 7 6 5 4 3 2 1

Contents

Preface

The International Symposium on Advanced Processing and Manufacturing Technologies for Structural and Multifunction Materials and Systems (APMT) was held during the 31st International Conference on Advanced Ceramics and Composites, in Daytona Beach, FL, January 21-26, 2007. The aim of this symposium was to discuss global advances in the research and development of advanced processing and manufacturing technologies for a wide variety of non-oxide and oxide based structural ceramics, particulate and fiber reinforced composites, and multifunctional materials. A total of 52 papers, including invited talks, oral presentations and posters, were presented from 10 countries (USA, Japan, Germany, Slovenia, Korea, India, Ireland, Turkey, Belgium, and the United Kingdom). The speakers represented universities, industry and research laboratories.

This volume contains 20 peer reviewed (invited and contributed) papers covering the latest developments in processing and manufacturing technologies, including binder and slurry technologies, novel forming and sintering technologies, smart processing/micro-fabrication, thin-film processing, large scale/complicated shape processing, and machining technologies. These papers discuss the most important aspects necessary for understanding and further development of processing and manufacturing of ceramic materials and systems.

The editors wish to extend their gratitude and appreciation to all the authors for their cooperation and contributions, to all the participants, symposium co-organizers and session chairs for their time and efforts, and to all the reviewers for their valuable comments and suggestions. Financial support from the Ube Industries, Ltd., Japan, as well as the Engineering Ceramic Division of The American Ceramic Society is gratefully acknowledged. Thanks are due to staff of the Meetings and Publications departments of The American Ceramic Society for their invaluable assistance.

We hope that this issue will serve as a useful reference for the researchers and technologists working in the field of advanced processing and manufacturing of ceramic materials and systems.

TATSUKI OHJI
Advanced Manufacturing Research Institute
AIST, Nagoya, Japan

MRITYUNJAY SINGH
Ohio Aerospace Institute
Cleveland, OH, USA

Introduction

2007 represented another year of growth for the International Conference on Advanced Ceramics and Composites, held in Daytona Beach, Florida on January 21-26, 2007 and organized by the Engineering Ceramics Division (ECD) in conjunction with the Electronics Division (ED) of The American Ceramic Society (ACerS). This continued growth clearly demonstrates the meetings leadership role as a forum for dissemination and collaboration regarding ceramic materials. 2007 was also the first year that the meeting venue changed from Cocoa Beach, where it was originally held in 1977, to Daytona Beach so that more attendees and exhibitors could be accommodated. Although the thought of changing the venue created considerable angst for many regular attendees, the change was a great success with 1252 attendees from 42 countries. The leadership role in the venue change was played by Edgar Lara-Curzio and the ECD's Executive Committee, and the membership is indebted for their effort in establishing an excellent venue.

The 31st International Conference on Advanced Ceramics and Composites meeting hosted 740 presentations on topics ranging from ceramic nanomaterials to structural reliability of ceramic components, demonstrating the linkage between materials science developments at the atomic level and macro level structural applications. The conference was organized into the following symposia and focused sessions:

- Processing, Properties and Performance of Engineering Ceramics and Composites
- Advanced Ceramic Coatings for Structural, Environmental and Functional Applications
- Solid Oxide Fuel Cells (SOFC): Materials, Science and Technology
- Ceramic Armor
- Bioceramics and Biocomposites
- Thermoelectric Materials for Power Conversion Applications
- Nanostructured Materials and Nanotechnology: Development and Applications
- Advanced Processing and Manufacturing Technologies for Structural and Multifunctional Materials and Systems (APMT)

- Porous Ceramics: Novel Developments and Applications
- Advanced Dielectric, Piezoelectric and Ferroelectric Materials
- Transparent Electronic Ceramics
- Electroceramic Materials for Sensors
- Geopolymers

The papers that were submitted and accepted from the meeting after a peer review process were organized into 8 issues of the 2007 Ceramic Engineering & Science Proceedings (CESP); Volume 28, Issues 2-9, 2007 as outlined below:

- Mechanical Properties and Performance of Engineering Ceramics and Composites III, CESP Volume 28, Issue 2
- Advanced Ceramic Coatings and Interfaces II, CESP, Volume 28, Issue 3
- Advances in Solid Oxide Fuel Cells III, CESP, Volume 28, Issue 4
- Advances in Ceramic Armor III, CESP, Volume 28, Issue 5
- Nanostructured Materials and Nanotechnology, CESP, Volume 28, Issue 6
- Advanced Processing and Manufacturing Technologies for Structural and Multifunctional Materials, CESP, Volume 28, Issue 7
- Advances in Electronic Ceramics, CESP, Volume 28, Issue 8
- Developments in Porous, Biological and Geopolymer Ceramics, CESP, Volume 28, Issue 9

The organization of the Daytona Beach meeting and the publication of these proceedings were possible thanks to the professional staff of The American Ceramic Society and the tireless dedication of many Engineering Ceramics Division and Electronics Division members. We would especially like to express our sincere thanks to the symposia organizers, session chairs, presenters and conference attendees, for their efforts and enthusiastic participation in the vibrant and cutting-edge conference.

ACerS and the ECD invite you to attend the 32nd International Conference on Advanced Ceramics and Composites (http://www.ceramics.org/meetings/daytona2008) January 27 - February 1, 2008 in Daytona Beach, Florida.

JONATHAN SALEM AND DONGMING ZHU, Volume Editors
NASA Glenn Research Center
Cleveland, Ohio

SLURRY CHARACTERIZATION BY STRESS RELAXATION TEST FOR TAPE CASTING PROCESS

Takamasa Mori, Tomofumi Yamada, Tatsuya Tanaka, Junichiro Tsubaki
Department of Molecular Design and Engineering, Nagoya University
Furo-cho, Chikusa-ku, Nagoya, 464-8603, Japan

ABSTRACT

Slurries were evaluated by conventional apparent viscosity measurement and a constant rate filtration and stress relaxation test proposed in this study in order to identify the most influential property of a slurry on the crack formation in a green sheet during drying. We evaluated the packing ability of slurry from the constant rate filtration and stress relaxation rate of a cake from the stress relaxation test. Slurries were also tape cast and dried at room temperature and then cracks formed in green sheets were observed. Slurry properties were controlled by changing pH value of slurry or additive amount of the binder.

It was shown that there is not good correlation between the apparent viscosity, packing ability of slurry and crack formation in green sheet, while the number of cracks decreased with an increase in stress relaxation rate of the cake. Stress relaxation test proposed in this study could be useful for the prediction of crack formation regardless of slurry preparation method.

INTRODUCTION

Paper-thin, flexible green sheets of various ceramic compositions are produced by tape casting for manufacturing substrates of electronic conductors, resistors, and multilayer capacitors. This forming process has following advantages: (i) easy control of the thickness of green sheets, (ii) easy handling and processing of green sheets, and (iii) enable to laminate green sheets. However, shape distortion and/or cracking often occur during drying, causing defects of final products[1-4]. Because the volatility and toxicity of organic solvents used in slurries are harmful to the environment and health, and the solvents are difficult to recycle, the aqueous tape casting has been widely studied[5-23]. Water, as a solvent, has the advantage of being non-toxic, non-flammable, easily available and cheap. Nevertheless, aqueous tape casting is not still widely adopted by industry, because the probability of crack formation in green sheets is much higher as compared to using organic solvent.

Properties of green sheets strongly depend on slurry properties, however it is still unknown the relationship between slurry properties and crack formation during drying and the optimal slurry conditions are usually determined by trial and error[5-11]. In many researches concerning the slurry characterization for tape casting[12-19], the apparent viscosity of the slurry was used to evaluate the slurry property in an ordinary way, although it does not have a good correlation with the drying behavior of green sheets, such as crack formation and critical cracking thickness.

During drying, the drying stress acts in a green sheet because of the capillary force and initiates cracking[1,3,20-23]. Therefore, the following two factors of a slurry should affect crack formation.

First, if particles in a green sheet were packed very closely and particles could not move easily, the green sheet endured the drying stress, resulting in crack free green sheet. In this case we can predict crack formation from the density of the green sheet. It was reported that the density of green bodies is influenced by the packing ability of a slurry[24-28], therefore, the crack formation is expected to have a close relation to the packing ability of a slurry.

Secondly, if the drying stress dispersed and decreased quickly, cracking should not occur. In this case we can predict crack formation by measuring the stress relaxation rate of the green sheet.

The aims of this study were to identify the most influential property of a slurry on the drying behavior of a green sheet and to establish the general slurry characterization method for tape casting process. From the above viewpoint, the particle packing ability and stress relaxation rate were evaluated by the constant rate filtration and stress relaxation test[29].

EXPERIMENTAL PROCEDURE
Slurry preparation
We used two types of slurries: one was pH adjusted slurries and the other was binder containing slurries.

pH adjusted slurries
Slurries were prepared from alumina powder (AES-11E with an average particle size of 0.48μm, Sumitomo Chemical, Japan) and distilled water by ball-milling for 2 h using alumina balls in a polyethylene bottle. A solution of hydrochloric acid was used to control the pH values of slurries. The pH values of slurries were 3.1, 3.7, 3.8, 3.9, and 5.4 in this study. Initial solid concentration was 45 vol%. Prepared slurries were degassed by vacuum treatment and then used for slurry characterization and tape casting test.

Binder containing slurries
The raw material was also alumina powder and polycarboxylic ammoniun (D-305, Chukyo Yushi, Japan) was used as dispersant. Polyvinyl alcohols (PVA) and polyethyleneglycol (PEG) were used as binder and plasticizer, respectively. Initial solid concentration was 35 vol%. First alumina powder, distilled water and dispersant were mixing by ball-milling for 1 h. After ball-milling, prepared slurries were degassed and poured into a container. PEG was added gradually into the slurries under stirring and then PVA was added subsequently. Total amount of PVA and PEG was changed 4, 5, 6, and 7 vol%, while the ratio of PVA and PEG was 1:1 for all slurries.

Slurry characterization
Prepared slurries were characterized by apparent viscosity measurement and constant rate filtration and stress relaxation test. Apparent viscosity was measured by a Brookfield type viscometer (BK type, Tokimec inc., Japan) at a shear rate of 7.3 s[-1].

Figure 1 shows the schematic illustration of the experimental apparatus for the constant rate filtration and stress relaxation test. The prepared slurry was poured into the filtration equipment and was placed it on the mechanical testing machine (SDW-2000, Imada co., Japan).

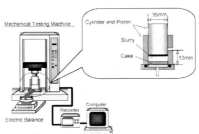

Fig.1   Schematic illustration of experimental apparatus for constant rate filtration and stress relaxation test of slurries

After that, filtration was started by moving the piston down at a constant speed of 0.14 mm⁻ min⁻¹. After filtration, the formed cake was compressed subsequently until the stress reached at 1 MPa, and then the valve at the bottom of the filtration equipment was closed and the strain of the cake was maintained constant. The stress relaxation behavior was observed until the stress became almost constant. For the binder containing slurries, the maximum stress was changed 3 MPa, because filtration was not completed when the stress reached at 1 MPa.

Tape casting
        Figure 2 shows the schematic illustration of the tape casting equipment used in this study.

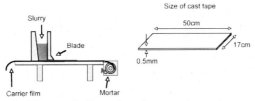

Fig.2 Schematic illustration of tape casting equipment

A gap of 0.5 mm under the blade and a casting speed of 145 mm/min were selected for all the casting test. The green sheet obtained from this experiment had 170 mm width and 500 mm length. The cast tape was drying at room temperature for at least 15 h. The career sheet with cast green sheet was fixed on a table in order to prevent distortion of it.

        After drying, we observed crack formation in green sheets by a digital camera. In order to evaluate cracks, we drew lines with a distance of 30 mm and counted the number of intersections between the cracks and the lines. The number of cracks was determined by dividing the number of intersections by the total length of all lines, which means the number of intersections between the cracks and the lines at unit length of the line.

RESULTS AND DISCUSSION
Slurry characterization
        The packing fraction of formed cake after filtration was calculated from the filtrate v and cake height L, using the following mass balance equation.

$$\Phi = \frac{v+L}{L}\phi_0 \quad (1)$$

The packing fractions of the cake are summarized in Table 1 for the pH adjusted slurries. Only for the slurry with a pH of 5.4, the packing fraction was relatively low, while the others had almost the same value.

Table 1 Packing fractions of cake obtained from pH adjusted slurries

Sample	Packing fraction [-]
pH3.1	0.49
pH3.7	0.49
pH3.8	0.49
pH3.9	0.48
pH5.4	0.46

Figure 3 shows the stress relaxation behavior of formed cake for pH adjusted slurries. In order to characterize these stress relaxation curves, the stress relaxation rate defined by following equation was introduced

$$(\text{Stress Relaxation Rate}) = \frac{(\text{Relaxation Stress})}{(\text{Relaxation Time})} \quad (2)$$

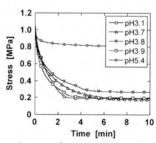

Fig.3 Time changes of stress acting on the formed cake

The stress relaxation rates were summarized in **Table 2** for pH adjusted slurries. The stress relaxation rate increased with a decrease in the pH value of slurry and the maximum stress relaxation rate was obtained for the slurry with a pH of 3.1.

The packing fractions of formed cake after filtration and stress relaxation rate were summarized in **Table 3** for binder containing slurries. We couldn't find clear tendency about packing fraction of cake for binder containing slurries. Roughly speaking, the stress relaxation increased with total amount of additives.

Table 2 Relaxation rate of pH adjusted slurries

Sample	Relaxation rate [kPa¥s⁻¹]
pH3.1	4.7
pH3.7	1.6
pH3.8	2.0
pH3.9	3.1
pH5.4	0.11

Table 3 Packing fraction and relaxation rate of binder containing slurries

Sample	Packing fraction [-]	Relaxation rate [kPa¥s⁻¹]
4	0.57	0.81
5	0.62	1.2
6	0.63	1.4
7	0.56	5.0
8	0.54	5.1

Characterization of green sheets

The number of cracks were summarized in **Tables 4** and **5**. For the pH adjusted slurries, we could obtain crack free green sheets from the slurry with a pH of 3.1. On the other hand the green sheets obtained from the slurry with a pH of 5.4 had many cracks and the number of cracks was quite large compared to those of others.

Table 4 Number of cracks in green sheet obtained from pH adjusted slurries

Sample	Number of cracks [m⁻¹]
pH3.1	0
pH3.7	7.0
pH3.8	2.2
pH3.9	2.1
pH5.4	57

Table 5 Number of cracks in green sheet for binder containing slurries

Sample	Number of cracks [m^{-1}]
4	21
5	6.1
6	2.9
7	0
8	0

For binder containing slurries, we could obtain crack free sheets from slurries with relatively large amount of additives. Unfortunately, degassing was difficult for slurries with large amount of additives because the viscosity of the slurry was high, resulting in very small defects at the surface of green sheet. However, we will eliminate these defects by improving the degassing method, therefore we neglected these defects for the calculation of the number of cracks.

RELATIONSHIP BETWEEN SLURRY PROPERTIES AND CRACK FORMATION
Apparent viscosity

Figure 4 shows the relationship between apparent viscosity of prepared slurries and the number of cracks. It was shown that there is not good correlation between apparent viscosity and the number of cracks and the apparent viscosity changed to almost three times lager even though the number of cracks remained zero. From these results, it is difficult to predict the crack formation in green sheet from apparent viscosity as shown in previous report[19].

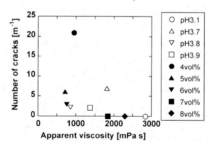

Fig.4 Relationship between number of cracks and apparent viscosity

Packing fraction of cake

Figures 5 and 6 show the relationship between the packing fraction of cake obtained from filtration and the number of cracks. Here, because the maximum applied pressure was different, we discuss the results of pH adjusted slurries and binder containing slurries, separately.

For pH adjusted slurries, roughly speaking, the number of cracks would decrease with an increase in packing fraction of cake. However, there was little difference between the packing fractions of the slurries with a pH of 3.1 (crack free) 3.8, and 3.9 (a few cracks). This means that we can not predict the crack formation strictly from the packing fraction of cake for pH adjusted slurries.

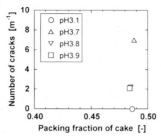

Fig.5 Relationship between number of cracks and packing fraction of formed cake for pH
        adjusted slurries

For binder containing slurries, there was no clear relation between the packing fraction of cake obtained by filtration and the number of cracks.

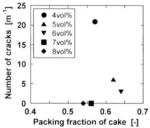

Fig. 6 Relationship between number of cracks and packing fraction of formed cake for biner
        containing slurries

In previous reports[29,30], some researchers discussed the relation of the crack formation and packing ability of slurry, however the results were quite different; some said that the films obtained from flocculate slurry have few cracks because the flocculate slurry made a film with loose packing of particles, namely lowering the capillary force during drying, while the other said that the coating cast from more flocculate slurry cracked earlier. From above discussion, the packing fraction of cake should not be the main factor affecting on crack formation in green sheets.

Stress relaxation rate
        Figures 7 and 8 show the relationship between the stress relaxation rate and the number of cracks for pH adjusted slurries and binder containing slurries, respectively. It was found that the number of cracks decreased with an increase of the stress relaxation rate for both slurries. In

addition there was a clear difference between slurries making crack free green sheets and green sheets with some cracks, while it was hard to distinguish them from the apparent viscosity of slurry or packing fraction of cake. These results suggest that the drying stress should be released quickly in the case of the slurry with higher stress relaxation rate, resulting in crack free green sheet, as mentioned in the introduction of this paper. After all, the most influential factor on the crack formation in green sheets must be the stress relaxation rate and the stress relaxation test can be useful to predict crack formation during drying regardless of slurry preparation method.

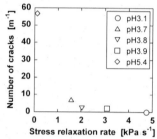

Fig.7 Relationship between number of cracks and relaxation rate for pH adjusted slurries

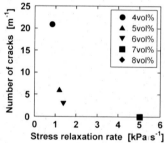

Fig.8 Relationship between number of cracks and relaxation rate for binder containing slurries

CONCLUSIONS

In order to identify the most influential property of a slurry on the drying behavior of a green sheet, some slurry characterizations including novel stress relaxation test and tape casting were conducted. It was shown that the number of cracks decreased with an increase in the stress relaxation rate and the stress relaxation rate had the closest relation to the number of cracks among the all slurry properties measured in this study. The stress relaxation test can be useful to predict crack formation during drying regardless of slurry preparation method such as pH adjusted or binder containing slurries.

ACKNOWLEDGEMENT
The authors would like to acknowledge the financial support of Hosokawa Powder Technology Foundation, Japan.
Noritake company ltd. is acknowledged for providing the apparatus for tape casting.

NOMENCLATURE

Φ	: packing fraction of the cake	(-)
ϕ_0	: initial solid volumetric concentration	(-)
L	: cake height	(m)
v	: volume of filtrate in unit area	(m)

REFERENCES

1. James S. Reed, Principles of Ceramics Processing, pp. 525-539 (1995)
2. N. Mizutani, T. Kimura, Y. Ozaki and T. Yamaguchi, Ceramic Processing, pp. 119-121 (1987)
3. George W. Scherer, Theory of Drying, J. Am. Ceram. Soc., 73(1), 3-14 (1990)
4. R. Misra, A. J. Barker and J. East, Controlled drying to enhance properties of technical ceramics, *Chem. Eng. Journal*, **86**, 111-116 (2002)
5. M. P. Albano and L. B. Garrido, Aqueous tape casting of yttria stabilized zirconia, Mater. Sci. Eng. A, 420, 171-178 (2006)
6. J. Gurauskis, A.J. Sanchez-Herencia and C. Baudin, Joining green ceramic tapes made from water-based slurries by applying low pressures st ambient temperature, *J. Euro. Ceram. Soc.*, **25**, 3403-3411 (2005)
7. S. Mei, J. Yang, X. Xu, S. Quaresma, S. Agathopoulos and J. M. Ferreira, Aqueous tape casting processing of low dielectric constant cordierite-based glass-ceramics –selection of binder, *J. Euro. Ceram. Soc.*, **26**, 67-71 (2004)
8. A. Navarro, J. R. Alcock and R. W. Whatmore, Aqueous colloidal processing and green sheet properties of lead zirconate titanate (PZT) ceramics made by tape casting, *J. Euro Ceram. Soc.*, **24**, 1073-1076 (2004)
9. F. Snijkers, A. Wilde, S. Mullens and J. Luyten, Aqueous tape casting of yttria stabilized zirconia using natural product binder, *J. Euro. Ceram. Soc.*, **24**, 1107-1110 (2004)
10. A. Kristoffersson, E. Roncari and C. Galassi, Comparison of different binders for water-based tape casting of alumina, *J. Euro. Ceram. Soc.*, **18**, 2123-2131 (1998)
11. M. Descamps, G. Ringuet, D. Leger and B. Thierry, Tape-casting: Relationship between organic constituents and the physical and mechanical properties of tapes, *J. Euro. Ceram. Soc.*, **15**, 357-362 (1995)
12. M. P. Albano and L. B. Garrido, Influence of the slip composition on the aqueous processing and properties of yttria stabilized zirconia green tapes, *Ceram. Inter.*, **32**, 567-574 (2006)
13. Y. L. Song, X. L. Liu, J. Q. Zhang, X. Y. Zou and J. F. Chen, Rheological properties of nanosized barium titanate prepared by HGRP for aqueous tape casting, *Powder Technol.*, **155**, 26-32 (2005)
14. S. M. Olhero and J. M. Ferreira, Rheological characterization of water-based AlN slurries for the tape casting process, *J. Mater. Proc. Technol.*, **169**, 206-213 (2005)
15. Z. Jingxian, J. Dongliang, L. Weisensel and P. Greil, Deflocculants for tape casting of TiO_2 slurries, *J. Euro. Ceram. Soc.*, **24**, 2259-2265 (2004)

16. A. Seal, D. Chattopadhyay, A. D. Sharma, A. Sen and H. S. Maiti, Influence of ambient temperature on the rheological properties of alumina tape casting slurry, *J. Euro. Ceram. Soc.*, **24**, 2275-2283 (2004)

17. Z. Yuping, J. Dongliang and P. Greil, Tape casting of aqueous Al_2O_3 slurries, *J. Euro. Ceram. Soc.*, **20**, 1691-1697 (2000)

18. A. Kristoffersson, R. Lapasin and C. Galassi, Study of interactions between polyelectrolyte dispersants, alumina and latex binders by rheological characterization, *J. Euro. Ceram. Soc.*, **18**, 2133-2140 (1998)

19. R. Greenwood, E. Roncari and C. Galassi, Preparation of concentrated aqueous alumina suspension for casting, *J. Euro. Ceram. Soc.*, **17**, 1393-1401 (1997)

20. J. Kiennemann, T. Chartier, C. Pagnoux, J. F. Baumard, M. Huger and J. M. Lamerant, Drying mechanisms and stress development in aqueous alumina tape casting, *J. Euro. Ceram. Soc.*, **25**, 1551-1564 (2005)

21. P. Wedin, J. A. Lewis and L. Bergstrom, Soluble organic additive effects on stress development during drying of calcium carbonate suspensions, *J. Colloid Interface Sci.*, **290**, 134-144 (2005)

22. C. J. Martinez and J. A. Lewis, Rheological, structural, and stress evolution of aqueous Al_2O_3 : latex tape-cast layers, *J. Am. Ceram. Soc.*, **85**, 2409-2416 (2002)

23. J. A. Lewis, K. A. Blackman and A. L. Ogden, Rheological property and stress development during drying of tape-cast ceramic layers, *J. Am. Ceram. Soc.*, **79**, 3225-3234 (1996)

24. T. Mori, K. Kuno, M. Ito, J. Tsubaki and T. Sakurai "Slurry Characterization by Hydrostatic Pressure Measurement –Analysis Based on Apparent Weight Flux Ratio-" *Advanced Powder Technol., Japan*, **17**, 319-332 (2006)

25. T. Mori, H. Kim, K. Ato and J. Tsubaki, Measuring packing fraction distribution in cake filtered under constant pressure and quantitative estimation from measured filtrate volume, *J. Ceram. Soc., Japan*, **114**, 318-322 (2006)

26. W. Sakamoto and S. Hirano, Processing of dielectric ceramics sheets using aqueous slurries, *Mater. Integration*, **19**, 25-33 (2006)

27. G. Bertrand, C, Filiatre, H. Mahdjoub, A. Foissy and C. Coddet, Influence of slurry characteristics on the morphology of spray-dried alumina powders, *J. Euro. Ceram. Soc.*, **23**, 263-271 (2003)

28. J. Tsubaki, M. Kato, M. Miyazawa, T. Kuma and H. Mori, The effects of the concentration of a polymer dispersant on apparent viscosity and sedimentation behavior of dense slurries, *Chem. Eng. Sci.*, **56**, 3021-3026 (2001)

29. T. Mori, T. Yamada, T. Tanaka, A. Katagiri and J. Tsubaki "Effects of Slurry Properties on the Crack Formation in Ceramics Green Sheet during Drying", *J. Ceram. Soc., Japan*, **114**, 823-828 (2006)

30. R. C. Chiu, T. J. Garino and M. J. Cima, Drying of granular ceramic films : I, effect of processing variables on cracking behavior, *J. Am. Ceram. Soc.*, **76**, 2257-2264 (1993)

31. W. Lan, X. Wang and P. Xiao, Agglomeration effect on drying of yttria-stabilised-zirconia slurry on a metal substrate, *J. Euro. Ceram. Soc.*, **26**, 3599-3606 (2006)

PREPARATION OF STABLE NANO-SIZED Al$_2$O$_3$ SLURRIES USING WET-JET MILLING

Toshihiro Isobe, Yuji Hotta, Kimiyasu Sato and Koji Watari
Advanced Manufacturing Research Institute, National Institute of Advanced Industrial Science and Technology (AIST)
2266-98 Anagahora, Shimo-Shidami, Moriyama-Ku,
Nagoya, Japan, 463-8560

ABSTRACT

The agglomerated alumina powders with average particle size of 43 nm were pulverized by wet-jet milling process and the particle size distribution and stability of the obtained slurries were evaluated. The alumina slurries were prepared from alumina powder, distilled water and dispersant. The solid content of the slurry was 30 vol. %. The particle size distribution of the sample before wet-jet milling showed the broad peak at 10 μm corresponding to the agglomerations of the alumina powders. The viscosity of this slurry was 253 mPa•s and rapidly increased with time. The particle size distribution of the slurry after wet-jet milling at 50 MPa showed the 0.2 and 10 μm peaks. These sizes were larger than primary particle size and were originated from the agglomeration of the alumina powder. On the other hand, the powder after wet-jet milling at > 150 MPa was almost of dispersed particle size with a small amount of the agglomerations. These agglomerations were completely dispersed by > 5 passes of wet-jet milling. The viscosity of the obtained slurry was 16 times as low as that of the slurry before wet-jet milling and was stable for a long time.

INTRODUCTION

Nano-sized ceramic particles are widely used as polishing agents, fillers for sealant, films and paints, additives for conductive pastes and electronic components, photocatalysis and inorganic pigment. The nano particles should be mono-dispersed for these applications, however it is also known that the dispersion of nano-sized particles is difficult because of high agglutinability.

For break and pulverization of ceramic powders, ball milling and beads milling methods are available [1-4]. Fadhel et al. [1] have investigated the dispersion of TiO$_2$ powder by the beads milling method. The average particle size after the beads milling was about 0.15 μm. Qiu et al. [2] have prepared the AlN slurry by the beads milling method with ZrO$_2$ media (diameter: 100 μmφ). The average particle size after beads milling was about 0.11 μm after 90 min of grinding. Tanaka et al. [3] have carried out the ball milling and the beads milling for grinding alumina powder and the particle sizes of the obtained powders by both methods were > 0.1 μm. They reported that the particle size of the milled samples was reduced from the original before milling. The size reduction of ceramic particles during ball milling was calculated by particle element method (PEM) [4]. This result indicated that the size reduction rate depends on milling energy and these conventional milling methods were not adapted for nanoparticles.

Thus, these methods are preferable for submicron-sized particles however an appropriate method to pulverize agglomerations of nanoparticles is not yet available. Wet-jet milling method is a newer milling technique of the ceramic powders by collisions between the particles, and is preferable to homogeneously disperse the ceramic particles in the slurry. Monodispersion of the submicron-sized alumina particles by this method has been also reported [5-7]. The milling energy of the conventional methods was calculated from a container size, media size and amount and rotation speed. That of wet-jet milling method was, on the other hand, calculated from the pressure (velocity) of slurry in the collision region (jet milling pressure). Thus, this method is also available to the nano-sized particles by high jet milling pressure. In this study, the relationship between the jet milling pressure and flocculation size was evaluated and the stable nano-sized alumina slurries were prepared.

EXPERIMENTAL PROCEDURE

Nano-sized γ-alumina (Nano Tek®, C. I. Kasei, Japan) was mixed with 70 vol. % distilled water and 3.2 mass% poly (acrylic acid) (Aron A-6114, Toagosei, Japan) as a dispersant using a planetary homogenizer [8] (AR-250, Thinky, Japan) for 10 min. The mixture was pulverized by a laboratory scale wet-jet milling system (PRE03-20-SP, Genus, Japan). The jet milling pressure was 50 – 200 MPa and the wet-jet milling was repeated 1 – 5 times. The viscosity and stability of the suspension were measured by Vibro-viscometer (SV-10, A&D, Japan) at 25°C. The particle size distribution of the alumina particles in the suspension was estimated by laser diffraction method (La-920, Horiba, Japan).

RESULTS AND DISCUSSION
Characterization of the slurry before wet- jet milling

Figure 1 shows the apparent viscosity of the slurry before wet-jet milling. The initial viscosity was 253 mPa•s and the viscosity increased with time. It is considered that the increase

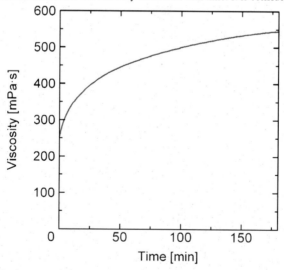

Fig. 1 Apparent viscosity of the slurry before wet-jet milling

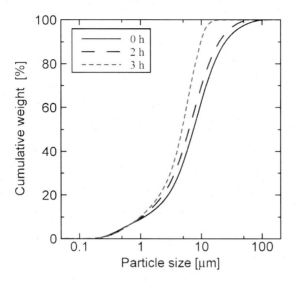

Fig. 2 Particle size distribution of the slurries before wet-jet milling

of the viscosity is due to re-flocculation of the alumina particles and the flocculated slurry shows thixotropic properties[9]. Figure 2 shows the particle size distribution of the slurry containing 30 vol. % alumina before wet-jet milling. To observe the re-flocculation of powders, those of the slurries after sitting for 2 and 3 h were also measured. The slurry before wet-jet milling shows with peak at 10 μm, which is larger than primary particle size (43 nm). And the particle size distribution of the slurries after sitting for 2 and 3 h shows decrease of the large particles (> 5 μm) and slight decrease of the small particles (< 0.5 μm). These deceases depend on setting time and indicate that the particles agglomerated and sedimented. This result shows a similar tendency to viscosity measurement.

Pulverization by wet-jet milling

Figure 3 shows the particle size distribution of the slurries after wet-jet milling at various milling pressure. The slurry treated with 50 MPa shows the 0.2 and 10 μm peaks, which were thought to be secondary and tertiary flocculation, respectively. The tertiary flocculation was pulverized at > 100 MPa of jet milling pressure. However, the particle size distribution of the slurries after wet-jet milling at > 150 MPa still showed a small amount of the agglomerations at about 1 μm. These agglomerations were completely crushed by repeating the wet-jet milling > 4 times. On the other hand, the secondary particle size decreased with increasing jet milling pressure. Figure 4 shows the relationship between jet milling pressure and mean size of the secondary flocculation. The secondary particle size decreased with increasing jet milling pressure but was satiated at 0.11 μm. This size is slightly larger than primary particle size. By contrast, the particle sizes of the slurries containing < 10 % powders were about 43 nm

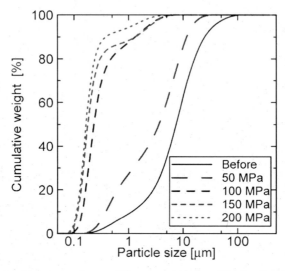

Fig. 3 Relationship between the jet milling pressure and particle size distribution of the slurries.

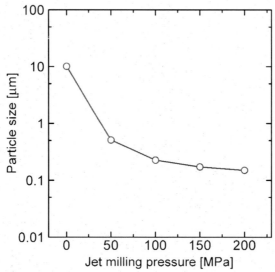

Fig. 4 Relationship between jet milling pressure and mean size of the obtained agglomerating powders.

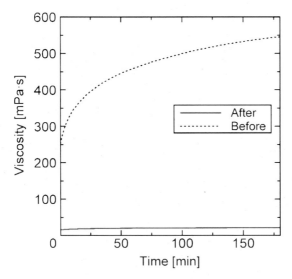

Fig. 5 Apparent viscosity of the slurry after wet-jet milling

corresponding to the primary particle size of the raw γ-alumina powder. Therefore, it is considered that this satiation may be due to a limitation of the dispersion of nano-sized particles in the concentrated slurries.

Stability of the obtained slurries

Figure 5 shows the viscosity of the slurry after wet-jet milling at 200 MPa. To comparison, that of the slurry before jet milling is shown in Fig. 5. The initial viscosity was 15.5 mPa•s, which was 16 times as low as that of the slurry before milling. The viscosity of the obtained slurry was stable for a long time. It is considered that thixotropic properties of the slurries significantly fell. In general, non-newtonian fluid, such as thixotropic and dilatant fluid, is not preferable from view point of the production efficiency. To address this problem, it is necessary to fall the thixotropic properties, however, an appropriate method for nanoparticles have not been reported. This wet-jet milling method may become one of the most suitable methods to lower the thixotropic properties.

Figure 6 shows the particle size distribution of the present slurries after standing for 0 – 24 h. The particle size distributions of the slurries after sitting were in excellent agreement with the original distribution (0 h), as expected. This result is indeed proof of the stability of the slurries.

SUMMARY

The agglomerated nano-sized alumina powders were pulverized by wet-jet milling method. The particle size distribution showed that the sample before wet-jet milling have a broad

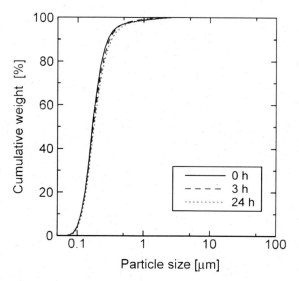

Fig. 6 Particle size distribution of the wet-jet milled slurries at (a) 200 MPa, repeating 5 times after 0 – 48 h

peak at 10 µm corresponding to the agglomerations. The viscosity of this slurry was 253 mPa•s and increased with time. The slurry showed thixotropic properties. On the other hand, the particle size distribution of the slurry after wet-jet milling at 50 MPa showed the 0.2 and 10µm peaks. The secondary particle size decreased with increasing jet milling pressure, and the samples after wet-jet milling at 200 MPa showed peak at 0.1 µm with a small amount of the tertiary agglomeration. These agglomerations were completely dispersed by repeating the wet-jet milling > 4 times. The viscosity of the obtained slurry was 16 times as low as that of the slurry before wet-jet milling. These results suggested that the dispersion and stability of the nano-sized particles depended on the collision energy and collision frequency. And the wet-jet milled slurry was stable for a long time and its thixotropic properties significantly fell.

REFERENCES

[1]H. Bel Fadhel, C. Frances, A. Mamourian, "Investigations on Ultra-fine Grinding of Titanium Dioxide in a Stirred Media Mill", *Powder Tech.*, **105**, 362-373 (1999).
[2]J.Y. Qiu, Yuji Hotta, Kimiyasu Sato and Koji Watari, "Fabrication of Fine AlN Particles by Pulverizing with Very Small ZrO₂ Beads", *J. Am. Ceram. Soc.*, **88**, 1676-1679 (2005)
[3]S. Tanaka, Z. Kato, N. Uchida, K. Uematsu, "Direct Observation of Aggregates and Agglomerates in Alumina Granules", *Powder Tech.*, **129**. 153-155 (2003).
[4]J. Kano, H. Mio, F. Saito and M. Tanjo, "Estimation of Size Reduction rate of Gibbsite in Tumbling Mills with Different Diameters by Computer Simulation", *J. Chem. Eng. Japan*, **32**, 747-751 (1999).

[5]N. Omura, Y. Hotta, K. Sato, Y. Kinemuchi, S. Kume and K. Watari, "Characterization of Al$_2$O$_3$ Slurries Prepared by Wet-jet Milling", *J. Ceram. Soc. Japan*, **113**, 491-494 (2005)

[6]N. Omura, Y. Hotta, K. Sato, Y. Kinemuchi, S. Kume and K. Watari, "Slip Casting of Al$_2$O$_3$ Slurries Prepared by Wet-jet Milling", *J. Ceram. Soc. Japan*, **113**, 495-497 (2005)

[7]N. Omura, Y. Hotta, K. Sato, Y. Kinemuchi, S. Kume, K. Watari, "Fabrication of Stable Al$_2$O$_3$ Slurries and Dense Green Bodies Using Wet-jet Milling", *J. Am. Ceram. Soc.*, **89**, 2738-2743 (2006)

[8] T. Isobe, Y. Kameshima, A. Nakajima and K. Okada, "Preparation and Properties of Porous Alumina Ceramics with Uni-directionally Oriented Pores by Extrusion Method Using a Plastic Substance as a Pore Former", *J. Eur. Ceram. Soc.*, **27**, 61-66 (2007)

[9] J.C. Chang, F.F. Lange and D.S. Pearson, "Viscosity and Yield Stress of Alumina Slurries Containing Large Concentrations of Electrolyte", *J. Am. Ceram. Soc.*, **77**, 19-26 (1994)

DESIGN OF MOLD MATERIALS FOR ENCAPSULATING SEMICONDUCTORS USING EPOXY COMPOUNDS

Satoshi Kitaoka and Naoki Kawashima
Japan Fine Ceramics Center
2-4-1 Mutsuno, Atsuta-ku
Nagoya, 456-8587, JAPAN

Keiji Maeda, Takaki Kuno and Yoshinori Noguchi
TOWA Corporation
5 Kamichoshi-cho, Kamitoba, Minami-ku
Kyoto, 601-8105, JAPAN

ABSTRACT

Mold materials have been found which have excellent releasabilities. This was achieved by inhibiting mold adhesion caused by acid-base interactions between epoxy molding compounds (EMCs) (used for encapsulating semiconductors) and hydroxyl groups on the mold surface. Some rare-earth oxides, which have isoelectric points of the surface (IEPS) in the range 9 to 10, have excellent releasability from EMCs. In particular, the releasing force of Y_2O_3 is 10 times smaller than that of conventional mold materials for alloy tool steels. It has thus received considerable attention for use as a mold material. It was found that the releasability depends on the arithmetic difference between the IEPS of the mold material and the dissociation constant of the EMC species.

Key words: mold, isoelectric point, adhesion, acid-base interaction

INTRODUCTION

Epoxy molding compounds (EMCs) are used for encapsulating semiconductors and are designed to adhere strongly to die surfaces and/or leadframes. However, the releasing force between the EMC and the encapsulation mold surface tends to be high. Therefore, the mold surface has to be cleaned frequently to remove the EMC adhered on the surface and is simultaneously coated with a mold lubricant[1]. This means serious reduction of productivity for integrated circuit (IC) packaging.

Furthermore, if the releasing force is uneven when the IC packages are ejected, there is a high probability of the die being damaged[2]. These adhesion problems are becoming more critical because EMCs having higher adhesions are being used to satisfy strict reliability requirements. It is thus highly desirable for mold materials to have excellent mold releasing properties for EMCs.

The surface of a mold is generally covered with hydroxyl groups due to the reaction between ambient water vapor and the oxide layer that forms on metallic molds. The stability of these hydroxyl groups is strongly dependent on the pH of solutions in contact with them[3,4]. If the pH of the solution is the same as the isoelectric point of the surface (IEPS) of the oxide, these hydroxyl groups will be undissociated. When the pH is less than the IEPS, the surface will acquire a positive charge,

$$- MeOH + H^+ \rightleftharpoons - MeOH_2^+. \tag{1}$$

If, on the other hand, the pH is greater than the IEPS, the surface will acquire a negative charge,

$$- MeOH \rightleftarrows - MeO^- + H^+. \tag{2}$$

The polarity of the surface charge affects the adsorption of polar organic compounds onto adherend materials. Bolger[3] proposed that the relative magnitude of the adsorption between an oxide and a polar organic compound can be estimated using the arithmetic difference between the IEPS of the oxide and the dissociation constant of the organic compound. Since EMC contains various kinds of polar organic compounds including epoxies, hardeners, accelerators and silane coupling agents, etc., Bolger's adhesion theory is expected to be applicable and should be useful for designing mold materials having excellent releasabilities for EMCs.

In this study, the adhesion strengths of various oxides were evaluated for conventional EMCs, and the materials having excellent releasabilities were determined. A criterion for designing mold materials having good releasabilities was proposed based on Bolger's adhesion theory.

EXPERIMENTAL PROCEDURES

Since EMCs are designed to adhere strongly to die surfaces and to the SiO_2 filler in IC packages, it is easy to conceive that acidic oxides such as SiO_2 and Ta_2O_5 will adhere to EMCs. In this study, two amphoteric oxides (Al_2O_3 and ZrO_2 (YSZ) with 4 mol% of Y_2O_3) and six basic oxides (MgO, Sc_2O_3, Y_2O_3, Yb_2O_3, Sm_2O_3 and Er_2O_3) were used in addition to alloy tool steel (SKD-11), a conventional mold material. These blocks were cut into cylinders having diameters of 13 mm and lengths of 10 mm. The adherend surfaces of the specimens were given mirror finishes to limit adhesion to EMC due to anchor effect as much as possible. These specimens were ultrasonic cleaned in acetone and were dried at 393 K for 24 hrs. The cleaned specimens were adhered to biphenyl type epoxy (a conventional EMC) by thermal curing at 443 K under a pressure of 10 MPa for 180 s. These conditions were selected in order to simulate actual encapsulation of semiconductors with the EMC. Cylindrical cured joints having diameters of 13 mm and lengths of 20 mm were obtained. The joint was set in a uniaxial tensile tester as shown in Fig.1. The adhesion strengths of the joints were measured at room temperature using a cross-head speed of 0.5 mm/min. The adhesive fracture surfaces of the specimens were analyzed by X-ray photoelectron spectroscopy (XPS).

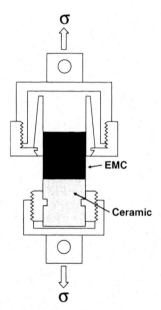

Figure 1 Schematic of the tensile test specimen showing a ceramic bonded with EMC.

RESULTS AND DISCUSSION

Figure 2 shows the adhesion strength of the EMC as a function of the IEPS of the ceramics. The IEPS was determined using Parks' equation[5] which depends on the valence and the radius of the cation. The IEPS of YSZ was calculated by distributing the values of its constituent oxides (ZrO_2 and Y_2O_3) with their molar ratios. The IEPS of SKD-11 is the same as that of Fe_2O_3 since it was assumed that its surface is entirely covered with a Fe_2O_3 layer due to oxidation. The adhesion strength of the non-rare-earth oxides which have IEPS below 8 and above 12 is about 10 MPa and these oxides were extremely strongly adhered to the EMC. By contrast, the adhesion strength of the rare-earth oxides, which have IEPS in the range between 9 and 10, is clearly smaller than those of YSZ, Al_2O_3 and MgO. In particular, the releasing force of Y_2O_3 is 10 times smaller than that of SKD-11. It thus has considerable potential as a mold material.

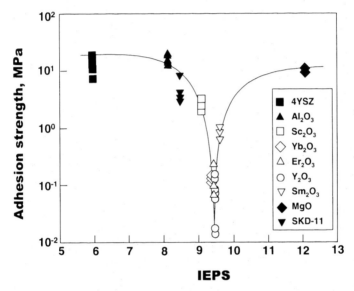

Figure 2 Adhesion strength of EMC as a function of IEPS of the ceramics.

Figure 3 shows optical micrographs of the adhesive fracture surfaces of the ceramics. The black regions in the micrographs indicate locations where the EMC was transferred to the ceramic surface. While there are many such regions distributed on the fracture surfaces of Al_2O_3 and MgO, there are few on those of Y_2O_3. The distribution of the locations where the EMC was transferred to the fracture surfaces of Al_2O_3 is significantly different from that of MgO, suggesting that the adhesion mechanism for the two oxides is different.

Near surface concentration gradients of adhesive fracture surfaces of the ceramics and EMC were evaluated by XPS angular profiling, where the electron takeoff angle θ with respect to the surface plane ranges from $15°$ to $70°$. Figure 4 shows the peak ratios of O/C and Al/C on the adhesive fracture surfaces of Al_2O_3 and EMC as a function of $\sin \theta$. The smaller $\sin \theta$ is, the smaller the detection depth is, and the closer the analyzed region is to the surface. The Al/C and O/C ratios of Al_2O_3 decrease with decreasing $\sin \theta$. When $\sin \theta$ is approximately 0.2, the Al/C and O/C ratios of Al_2O_3 become nearly zero and 0.1, respectively. However, the Al/C and O/C ratios for the EMC are independent of $\sin \theta$, being zero and 0.1, respectively. On the basis of this evidence, we can conclude that Al_2O_3 was not transferred to the fracture surface of the EMC. Since the O/C of the top surface of Al_2O_3 is approximately the same as that of the EMC, the surface of Al_2O_3 is coated with EMC and a crack propagates in the EMC. The adhesion strength of Al_2O_3 was the same as that of YSZ as Fig.2 shows. The fracture of the joint consisting of YSZ and the EMC also preferentially occurred in the EMC. By contrast, the adhesive fracture surfaces of Y_2O_3 and MgO were not coated with EMC. Since no peaks assigned to silicon 2P were detected for the surfaces of these oxides by the XPS angular profiling, the silane coupling agent that had been added to the EMC probably did not contribute to the adhesion of the EMC to these oxides.

Figure 5 shows the carbon 1s spectrum at a takeoff angle of 50 degrees obtained from the adhesive fracture surface of MgO. A peak assigned to CO_3 was observed in the spectrum of the fracture surface of MgO. It is well known that MgO reacts with CO_2 to produce hydroxy carbonate if water is present [6]. The carbonate had probably already formed on the adherend surface of MgO before the cured joints were produced.

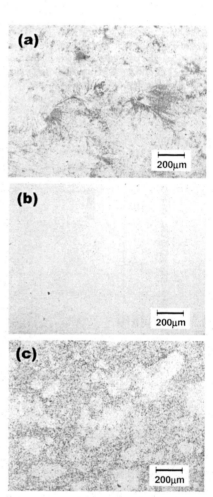

Figure 3 Optical micrographs of the adhesive fracture surfaces of the ceramics; (a) Al_2O_3, (b) Y_2O_3 and (c) MgO. The black regions in the micrographs indicate locations where the EMC transferred to the ceramic surfaces.

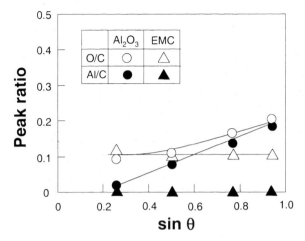

Figure 4 The peak ratios of O/C and Al/C on the adhesive fracture surfaces of Al$_2$O$_3$ and the EMC as a function of $sin\ \theta$, where θ is the takeoff angle with respect to the surface plane.

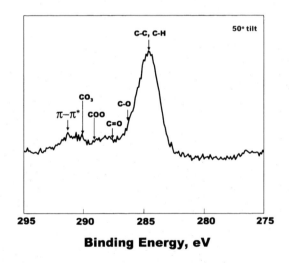

Figure 5 Carbon 1s spectrum at a takeoff angle of 50 degrees obtained from the adhesive fracture surface of MgO.

The π-π^* peak[7] assigned to an aromatic ring was also observed in the spectrum of the fracture surface of MgO. This peak is thought to be due to the phenol resin added as hardener to the EMC since it is weakly acidic and is a polar organic compound with an aromatic ring.

The surface of basic oxide interacts with acid organic compounds according to the following reaction,

$$\text{- Me}\overset{\cdot}{\text{O}}\text{H} + \text{HXR} \longrightarrow \text{- MeO}\cdots\overset{\overset{\text{H}}{|}}{\text{HXR}} \rightleftarrows \text{- MeOH}_2^+ \cdots \text{X}^-\text{R} \tag{3}$$

where X is oxygen and R changes from an aliphatic group to an aromatic group to a carboxylic group. In the case when EMC also contains basic organic compounds (for example, amine which acts as an accelerator) the surface of the acid oxide interacts with basic organic compounds as follows.

$$\text{- MeOH} + \text{XR} \longrightarrow \text{- MeOH}\cdots\text{XR} \rightleftarrows \text{- MeO}^- \cdots \text{H}^+\text{XR} \tag{4}$$

According to Bolgar's adhesion theory[3], the equilibrium constants for equations (3) and (4) are expressed by:

$$\log \frac{[\text{MeOH}_2^+]\cdot[\text{X}^-\text{R}]}{[\text{MeOH}]\cdot[\text{HXR}]} = \text{IEPS} - pK_{A(a)} \tag{5}$$

$$\log \frac{[\text{MeO}^-]\cdot[\text{H}^+\text{XR}]}{[\text{MeOH}]\cdot[\text{XR}]} = pK_{A(b)} - \text{IEPS} \tag{6}$$

where $pK_{A(a)}$ and $pK_{A(b)}$ are dissociation constants of the acid compounds and conjugate acid of the basic compounds, respectively.

The progress of the reactions of equations (3) and (4) leads to adsorption of the organic compounds on the oxide surfaces, resulting in an increase in the adhesion force. In other words, these reactions do not progress much on the surfaces of mold materials having very high releasabilities. The criterion for selecting releasable mold materials is given by:

$$pK_{A(b)} < \text{IEPS} < pK_{A(a)} \tag{7}$$

The rare-earth oxides, which have IEPS in the range 9 to 10, had excellent releasabilities as shown in Fig.2. The IEPS of the oxides probably satisfy this criterion. Namely, $pK_{A(a)}$ in equation (7) of the oxides is thought to be approximately equal to that (=10) of phenol. The basic organic compounds related to $pK_{A(b)}$ could not be identified, because the fracture surfaces of the oxides with an IEPS below 8 were coated with the EMC. Amine compounds used as accelerators may contribute to the strong adhesion.

CONCLUSION

Some rare-earth oxides have excellent releasabilities from EMCs. In particular, since Y_2O_3 has a releasing force that is 10 times smaller than that of the conventional metallic mold materials, it is receiving considerable interest for use as a mold material. The criterion for selecting mold materials having an excellent releasability is that their IEPS lies between the dissociation constant of the conjugate of basic organic compounds and that of acid organic compounds.

REFERENCES

[1] S.-J. Chang and S.-J. Hwang, "Design and Fabrication of an IC Encapsulation Mold Adhesion Force Tester", *IEEE Trans. on Electronics Packing Manufacturing*, **26**, 4, 281-285 (2003).

[2] M. Yoshii, Y. Mizutani and H. Shoji, "Evaluation Technologies on Moldability of Epoxy Molding Compounds for Encapsulation of Semiconductors", *Hitachi Chemical Technical Report*, **40**, 13-20 (2003).

[3] J. C. Bolger, "Acid Base Interactions Between Oxide Surface and Polar Organic Compounds", *Adhesion Aspects of Polymeric Coatings*, Ed., K. L. Mittal, Plenum Press, New York and London, 3-18 (1983).

[4] E. McCafferty and J. P. Wightman, "Determination of the Surface Isoelectric Point of Oxide Films on Metals by Contact Angle Titration", *J. Colloid and Interface Science*, **194**, 344-355 (1997).

[5] G. A. Parks, "The Isoelectric Points of Solid Oxides, Solid Hydroxides, and Aqueous Hydroxo Complex Systems", *Chem. Rev.*, **65**, 177-198 (1965).

[6] K. Yanagiuchi, "Effect of Decontamination on Metal Oxides by the Heating in the Air", *J. Surface Analysis*, **1**, 3, 395-401 (1995).

[7] D. H. Berry and A. Namkanisorn, "Fracture Toughness of a Silane Coupled Polymer-Metal Interface: Silane Concentration Effects", *J. Adhesion*, **81**, 347-370 (2005).

EVALUATION OF AN ENVIRONMENTALLY FRIENDLY PLASTICIZER FOR POLYVINYL BUTYRAL FOR USE IN TAPE CASTING

Richard E. Mistler[1], Ernest Bianchi[2], Bruce Wade[3], and Jeffrey Hurlbut[3]

[1]Richard E. Mistler, Inc.
Yardley, PA USA

[2]Maryland Ceramic & Steatite Co., Inc.
Bel Air, MD, USA

[3]Solutia, Inc.
Springfield, MA, USA

ABSTRACT
An environmentally friendly plasticizer for polyvinyl butyral has been evaluated and compared with the industry standard – butyl benzyl phthalate. Properties such as the effect on the glass transition temperature of polyvinyl butyral and its compatibility with several grades of polyvinyl butyral are discussed. A direct comparison of this plasticizer is made with the industry standard in side-by-side tape casting runs.

INTRODUCTION
The object of this investigation was to evaluate a more environmentally friendly plasticizer for use with polyvinyl butyral in tape casting formulations. The new plasticizer, triethylene glycol di-2-ethylhexanoate (S-2075), was compared directly with one of the most common plasticizers used in tape casting, butyl benzyl phthalate.

The paper is divided into two sections, one relating to the environmental and health comparisons between the S-2075 plasticizer and the butyl benzyl phthalate plasticizer and the direct comparison of their effects as additives to the polyvinyl butyral binder and the second which relates to a direct comparison of tape formulations using the two different plasticizers.

The comparisons made included: the effect of each plasticizer on the T_g, glass transition temperature, and the compatibility level for each of the most common grades of polyvinyl butyral, including Butvar® brands B-76, B-79, B-90, and B-98. The tape casting formulation comparisons included the following: slip viscosity, green bulk density (GBD), green oxide only density (GOOD), green tape thickness, tape drying characteristics, green tape character including flexibility and tackiness, and sintered properties such as fired bulk density (FBD) and shrinkage.

PART 1 - PLASTICIZER COMPARISONS:

ENVIRONMENTAL AND HEALTH CONSIDERATIONS

Butyl benzyl phthalate (BBP), commercially sold as Santicizer[R] 160, has two environmental and health issues: 1) reproduction toxicity and 2) dangerous for the environment. These issues impact mainly Europe, but spread in particular into Asia as producers there are becoming more and more concerned about imports into Europe. Specifically, BBP is classified according to its MSDS as "toxic to reproduction" Category 2 which is to be considered as toxic to reproduction to man. Another widely used phthalate plasticizer for tape casting, diethylhexyl phthalate (commonly called dioctyl phthalate), is also a Category 2 reprotoxic chemical. Phthalates, including BBP, are expected to be included on the "dangerous substances" or "substances to be avoided" industry chemical lists. BBP is also considered as very toxic to aquatic life and has a potential to bio-accumulate. This has less of an impact, but it compares poorly against S-2075 plasticizer. The data for S-2075 show that it is not toxic to aquatic life (LC50 and EC50) > 97 mg/l) and that it is biodegradable. Therefore, based on the currently available data, S-2075 has a better profile than BBP and other phthalate plasticizers.

PLASTICIZER EFFECTS ON POLYVINYL BUTYRAL:

Effect of Plasticizer Content on Tg
(Butvar® with Solutia S-2075 or Santicizer® 160 Plasticizer)

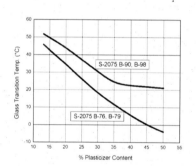 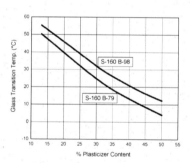

The glass transition temperature (Tg) profiles were determined by Dynamic Mechanical Analysis (DMA) with premixed Butvar® and plasticizer samples pressed into sheets. B-76 and B79 are 11-13% hydroxyl grades of PVB. B-90 and B-98 are 18-20% hydroxyl grades. The hydroxyl levels are expressed as weight percent polyvinyl alcohol.

The glass transition temperature (T$_g$) curves were determined by dynamic mechanical analysis (DMA) at a frequency of 1 Hz. Several glass transition measurements covering the plasticizer content range for each plotted curve were obtained. Samples of PVB and plasticizer were well-mixed in a Brabender Model D-51T and pressed into 0.040 inch thick sheets prior to T_g measurements.

During DMA testing, a material is subjected to a sinusoidal strain at a fixed frequency resulting in a sinusoidal stress which is at the same frequency but is out of phase with the applied strain. From this the complex modulus (E*) and the phase angle (δ) can be measured. The in-phase and out-of phase components of E* represent the Storage Modulus (E') and Loss Modulus (E''), respectively. The damping or dissipation factor (tan δ) is the ratio of the loss and storage moduli.

DMA is generally conducted over a wide temperature range in order to determine changes in E', E'' and tan δ. The primary glass-rubber transition (T_g) is characterized by a large decrease in E' and a maximum in both the E'' and Tan δ curves. The primary transition occurs in the amorphous regions of the polymer with the initiation of cooperative micro-Brownian motion of the molecular chains. The value of tan delta is sensitive to molecular mobility and is commonly used to determine the values of transition temperature in polymers.

The glass transition temperatures of PVB's containing a wide range of each plasticizer were determined using a Rheometrics Solids Analyzer (RSA-II) at a frequency of 1 Hz and a heating rate of 3°C/minute. Increasing plasticizer level is expected to result in a decrease in T_g. By plotting the T_g of each blend as a function of plasticizer level the level at which no further decrease in T_g can be determined. This level is taken to indicate the point at which the polymer can no longer take up any additional plasticizer.

Solutia S-2075 is found to have better compatibility and efficiency as measured by glass transition temperature than traditional Santicizer® S-160 butyl benzyl phthalate as well as other phthalate plasticizers, such as dioctyl phthalate evaluated in subsequent work, up to approximately 40% plasticizer content for 18-20% hydroxyl containing PVB. Up to 40% plasticizer content in PVB is more than sufficient to achieve a normal ambient temperature plasticization of 25°C. Solutia S-2075 is even more efficient than phthalate plasticizers, such as S-160 with low hydroxyl 10-13% type PVB, at all levels. S-2075 therefore can be used in lower concentrations to plasticize ceramic tapes.

PART 2 - TAPE CASTING FORMULATION COMPARISONS:

EXPERIMENTAL PROCEDURE

Two tape casting batches were prepared with everything exactly the same with the exception of the primary plasticizer. The ceramic batch which was selected for the comparison was a 94% aluminum oxide formulation which is commonly utilized in thick film or multilayered ceramic packages. The formulation for the control batch was as follows:

Part 1:

Aluminum Oxide [1]	54.70 weight %
Clay [2]	1.19 weight %
Talc [3]	3.57 weight %
Fish Oil [4]	1.19 weight %
Xylenes [5]	12.74 weight %
Ethyl Alcohol, 95% [6]	12.74 weight %

Part 2:

Butyl Benzyl Phthalate [7]	4.55 weight %
Polyalkylene Glycol [8]	4.55 weight %
Polyvinyl Butyral [9]	4.76 weight %

The formulation for the experimental batch was as follows:

Part 1:

Aluminum Oxide [1]	54.70 weight %
Clay [2]	1.19 weight %
Talc [3]	3.57 weight %
Fish Oil [4]	1.19 weight %
Xylenes [5]	12.74 weight %
Ethyl Alcohol, 95% [6]	12.74 weight %

Part 2:

Triethylene Glycol Di-2-Ethylhexanoatel [10]	4.55 weight %
Polyalkylene Glycol [8]	4.55 weight %
Polyvinyl Butyral [9]	4.76 weight %

The fish oil is a dispersant for the inorganic components. The butyl benzyl phthalate and triethylene glycol di-2-ethylhexanoate are considered Type I plasticizers for the polyvinyl butyral and the polyalkylene glycol is a Type II plasticizer for the tape. The polyvinyl butyral is the binder for the system and it provides the strength and backbone for the tape.

The procedure followed for each batch was identical and was as follows:

1. Add 2 Kg of 1 inch U.S. Stoneware Burundum 96% alumina grinding media to a size 1 Roalox mill jar. This is about 1/3 capacity.
2. Dissolve the fish oil in the xylenes and add to the mill jar.
3. Add the ethyl alcohol to the mill jar.
4. Add the powders to the mill jar; the alumina was dried at $> 100^0 C$ for 24 hours.
5. Dispersion mill at about 56 RPM for 24 hours on a jar roller.

6. Weigh and add the plasticizers to the mill jar.
7. Add the binder to the mill jar, stirring by hand to wet and mix the binder.
8. Mix for an additional 24 hours at about 56 RPM.
9. Pour the slip into HDPE containers.
10. Vacuum de-air for 8 minutes at 25 inches of Hg.
11. Measure the viscosity and temperature of the slip.

At this point the batches were ready for tape casting using the following casting parameters:

1. Blade Gap, two casts one at 0.030" and one at 0.050".
2. Ten (10) inch wide single doctor blade.
3. Carrier: Silicone Coated Mylar [11], 12 inch wide x 0.003 inch thick.
4. Casting Speed: 20 inches per minute.
5. Air flow on lowest setting with no heat.
6. No underbed heating.

After drying the following measurements were made on the green tape:

1. Thickness
2. Green Bulk Density

Samples for sintering were punched from the control and experimental tapes. The punched pieces, which were 1" x 1.5", had the long axis oriented in the casting direction and the short axis oriented in the cross-casting direction. This provided a good basis for comparison of the shrinkage during sintering in the casting direction and in the cross casting direction.

Two samples from each casting run were sintered in an electric furnace in an ambient air atmosphere. The samples were sandwiched between two porous alumina cover plates during sintering to prevent warping and to maintain flatness.

The temperature/time schedule for the sintering was as follows:

RT to 500°C @ 3°C per minute
500°C to 650°C @ 1°C per minute
Hold at 650°C for 0.5 hour
650°C to 1450°C @ 5°C per minute
1450°C to 1500°C @ 1°C per minute
Hold at 1500°C for 2 hours
Furnace Cool to RT

After sintering the samples were measured to determine the sintering shrinkage in the X, Y, and Z directions. Fired bulk density measurements were made using the Archimedes method by immersion in toluene.

RESULTS AND DISCUSSION

The viscosity measurements on the ready to cast slips were made using an RV-4 spindle at 20 RPM using a Brookfield Viscometer. The measurements were as follows:

Control Batch	4750 cP	at	23.5° C
	4500 cP	at	25.8° C
Experimental Batch	3650 cP	at	27.4° C

There were two measurements made on the control batch since the slip was divided into two portions for casting at two different doctor blade gap settings performed at different times during the day. As the room temperature increased the viscosity decreased as expected. There was only one measurement made for the experimental batch since only one cast was made with an adjustment of the doctor blade gap during the cast. It is obvious from the results that the viscosity was lower for the experimental batch. There is not enough data to determine whether this was due to the change in plasticizer or to the higher temperature at the time of the measurement. Certainly part of the viscosity reduction was due to the higher slip temperature. It appears, however, that almost a 1000 cP difference is not completely attributable to the higher temperature.

Table 2 includes the data and observations made on the tapes which resulted from the casting runs.

Table 2 – Green Tape Results

Property:	Control:	Experimental:
Thickness:		
Thin:	0.0102 -0.0144"	0.0095-0.0100"
Thick:	0.0161-0.0217"	0.0226-0.0242"
Green Bulk Density (GBD)	2.475 g/cc	2.56 g/cc
Green Oxide Only Density (GOOD)	1.975 g/cc	2.04 g/cc
Flexibility	High	Higher
Tackiness	None	Slight
Strength	Excellent	Excellent

The control batch of tape exhibited a typical curling at the edges of the cast which in turn caused some of the thickness variation observed. The thinner control tape exhibited less of this thickness variation. The curling of the experimental tape using the new plasticizer was not as severe and resulted in better thickness uniformity for both the thin and the thick tapes. A possible explanation for the better thickness uniformity with the experimental tape could be due to the slightly higher flexibility. It was also observed for the experimental lot that there was a faint residual film on the carrier and on the bottom side of the tape itself which is an indication that the plasticizer concentration may be slightly high. The tackiness observed is another indication of this. This ties in well with the T_g measurements which indicated that the S-2075 is a better plasticizer than the butyl benzyl phthalate. This leads to the conclusion that less plasticizer is needed to yield the same tape flexibility.

The green bulk density values are averages of several measurements on both the thin and thick tapes. The green oxide only density values are calculated from the GBD measurements and eliminate the organic content. Both the GBD and the GOOD values are significantly higher for the experimental tape. This is an indication that the actual packing density in the experimental tape is better. It is possible that the S-2075 is acting as a secondary dispersant in conjunction with the fish oil. Further experimental tests would have to be conducted to validate this.

The results for the sintering evaluation and comparison are presented in Table 3 as follows:

Table 3 - Sintering Results

Property:	Control:	Experimental:
Fired Bulk Density	3.618 g/cc	3.622 g/cc
Shrinkage:		
Along Tape	21%	20%
Across Tape	20%	20%
Thickness	10 – 14%	11- 15%

The results for the fired bulk density for the batches made with the standard butyl benzyl phthalate and the triethylene glycol di-2ethylhexanoate plasticizers are well within the experimental error limits of the measurements and can be considered equivalent. Likewise the fired shrinkage in the X, Y, and Z directions are also well within the experimental error limits. In addition all of the samples which were sintered were completely free of any defects.

SUMMARY AND CONCLUSIONS

Tape casting formulations for a standard 92% alumina ceramic have been evaluated using an environmentally friendly plasticizer. The results were compared with the industry standard plasticizer: butyl benzyl phthalate. The green tape properties including green bulk density, thickness uniformity, flexibility and strength were all found to be equal to or better than the standard. The final sintered density, shrinkage during sintering, and fired part quality also were equal to or better than the samples prepared using the industry standard. Based on these results it has been determined that the triethylene glycol di-2-ethylhexanoate Type 1 plasticizer can be used as a substitute for the butyl benzyl phthalate in tape casting formulations. As a substitute, Solutia S-2075 triethylene glycol di-2-ethylhexanoate is found to have better compatibility and efficiency with polyvinyl butyral as measured by glass transition temperature than butyl benzyl phthalate as well as other phthalate plasticizers.

1. Aluminum Oxide, A-16SG, Almatis, Leetsdale, PA 15056
2. EPK Kaolin, Zemex Industrial Minerals, Inc., Atlanta, GA 30338
3. Nytal® 400 Talc, R.T. Vanderbilt Co., Inc., Norwalk, CT 06856
4. Blown Menhaden Fish Oil, Grade Z-3, W.G. Smith, Inc., Cleveland, OH 44113
5. Xylenes, Reagent Grade
6. Ethyl Alcohol, 95% Denatured, Reagent Grade
7. Santicizer® 160, Ferro Corporation, Bridgeport, NJ 08014
8. UCON50HB2000, The Dow Chemical Co., Danbury, CT 06817
9. Butvar® B-98, Solutia Inc., St. Louis, MO 63166
10. S-2075, Solutia, Inc., St. Louis, MO 63166
11. G10JRM, Silicone Coated Mylar, Richard E. Mistler, Inc., Yardley, PA 19067.

MUTUAL LINKAGE OF PARTICLES IN CERAMIC GREEN BODIES THROUGH REACTIVE ORGANIC BINDERS

Kimiyasu Sato, Miyuki Kawai, Yuji Hotta, Takaaki Nagaoka, Koji Watari
National Institute of Advanced Industrial Science and Technology (AIST)
2266-98, Anagahora, Shimoshidami, Moriyama-ku
Nagoya, Japan

Cihangir Duran
Gebze Institute of Technology
P.K. 141, 41400
Gebze-Kocaeli, Turky

ABSTRACT
        Because the pyrolysis of organic substances can result in the emission of carbon dioxide or other hydrocarbon gases, a reduction in the use of organic binders is one aim of current ceramics industry. A novel ceramic-forming process was developed that requires considerably less organic binder than conventional techniques. The process involves immobilizing reactive molecules on the surfaces of the ceramic particles, which on subsequent irradiation with electromagnetic waves, form bridges that bind the entire particle assembly together. The chemical forces involved produce strong bonds, resulting in a significant reduction in the amount of organic binder that is required to maintain the shape of the ceramic green body. This method will help to decrease emission gases produced from pyrolysis of the binder.

INTRODUCTION

        In ceramic processing, organic substances are often used as binders, dispersants, plasticizers, or lubricating agents.[1,2] However, these must be removed before the sintering process by converting them into carbon dioxide and hydrocarbon gases, which are emitted into the environment. As a result of serious global environmental problems, all manufacturing industries are being forced to pay more attention to reduce the emission of pollutants. The current trend in the ceramic industry is to protect environment by reducing the amounts of organic additives that are used. In addition to environmental concerns, organic substances added to ceramic green bodies have to be removed, requiring extended heating times and leading to a decrease in manufacturing efficiency and an increase in energy consumption. Reducing the use of organic binders should also assist in circumventing such processing nuisances.

        Binders are traditionally employed in the shaping of ceramic materials because of their non-plastic nature. A conventional organic polymer binder functions by absorption onto the surfaces of ceramic particles that otherwise have insufficient mutual binding forces; however, a poor affinity between the binder molecules and the ceramic particle surfaces will result in phase separation and inhomogeneous partial segregation, which impair the function of the binder. The phase separation results in non-uniform microstructures in green bodies and may result in defects such as cracks and voids in sintered bodies.[3] Weak bonding by the binder and subsequent phase separation necessitate the use of disproportionately large amounts of organic binder. Though, shaping methods which do not rely on organic binders have been receiving attentions recently,[4-7]

35

these methods are based on the unique characteristic of specific ceramic powders such as hydraulic reactions. In cases where such specific schemes can not be employed due to material limitations, reduction of the binder amount is still required to prevail over the above-mentioned obstacles.

In the present study, we prepared ceramic green bodies using reactive organic thin-film layers which anchor covalently on the particle surfaces. Stronger bonding of the organic binders prevents phase separation between the ceramic particles and binder. The organic thin-films on the surface subsequently interact and act as a bridge for the linkage of the whole particle assembly (Fig. 1). Due to the use of such starving quantities of the binder molecules, the resultant green body contains only minimal amount of organic binder. In the forming process, ceramic particles should be able to merge mutually in an arbitrary stage. We investigated the utilization of two kinds of external stimuli (UV-light and microwave irradiations) as triggers for mutual binding reactions.

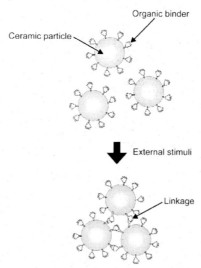

Figure 1. General idea of the novel organic binder system proposed in the present study.

PHOTOREACTIVE ORGANIC BINDER

The main characteristic of photoreactions via UV-light irradiation is that organic substances with specific structures can be selectively excited and connected. Photoreactive organic molecules were first anchored onto the surface of ceramic particles. The particles covered with a thin-film were then shaped and irradiated with UV-light to form mutual covalent linkages among each other.

Spherical silica particles (Fuso Chemical Co., Ltd., Osaka, Japan) were used as the ceramic phase. The number average size of the particles was 270 nm in diameter as observed by TEM. 3-aminopropyltriethoxysilane (3-APS; $NH_2C_3H_6Si(OC_2H_5)_3$, Shin-Etsu Chemical Co., Ltd., Tokyo, Japan) and 4-azidobenzoic acid ($N_3C_6H_4COOH$, Wako Pure Chemical Industries, Ltd., Osaka, Japan) were used as received. 200 ml of 3-APS aqueous solution (90 mM) was reacted with 10.0 g of silica particles under vigorous stirring for 4 hours. The resulting powdery material was washed by copious amounts of water and dried overnight at 105 °C. The dried powder was washed with water again to remove any unreacted 3-APS. This procedure resulted in silica particles whose surfaces are covered with amino groups (-NH$_2$) as shown in Scheme 1(a).[8-10] The silica particles treated with 3-APS will be designated as APS-silica hereafter. 4.5 g of APS-silica was dispersed in 300 ml of ethanol using intense agitation. 1.5 g of 4-azidobenzoic acid was added to the suspension with special care for shielding ambient light. The reactant was purified by centrifugation to remove excess 4-azidobenzoic acid molecules. After purification, the particles were heated to 180 °C for 6 hours under vacuum (1mmHg). A carboxyl group in 4-azidobenzoic acid and an amino group on the APS-silica surface form a secondary amide bonding with dehydration (Scheme. 1b).[11] The silica particles treated with 4-azidobenzoic acid will be designated as Az-silica hereafter.

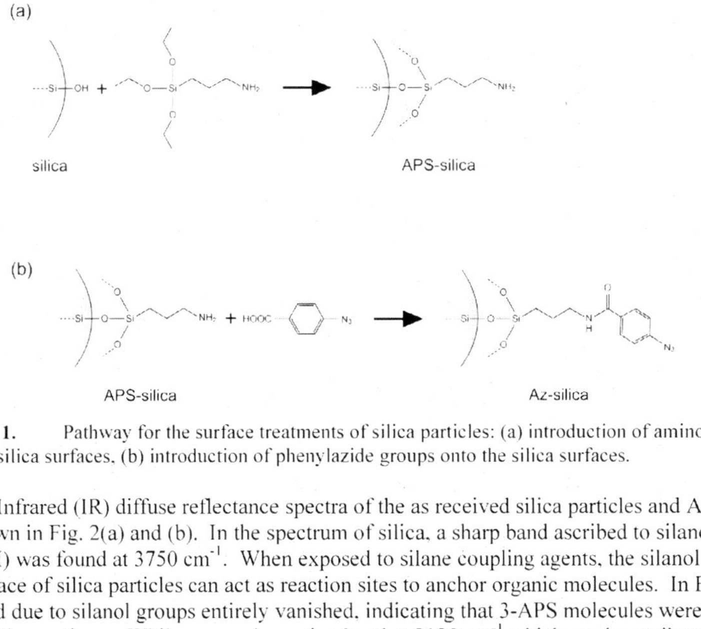

(a)

silica   APS-silica

(b)

APS-silica   Az-silica

Scheme 1. Pathway for the surface treatments of silica particles: (a) introduction of amino-groups into the silica surfaces, (b) introduction of phenylazide groups onto the silica surfaces.

Infrared (IR) diffuse reflectance spectra of the as received silica particles and Az-silica are shown in Fig. 2(a) and (b). In the spectrum of silica, a sharp band ascribed to silanol groups (-Si-OH) was found at 3750 cm^{-1}. When exposed to silane coupling agents, the silanol groups on the surface of silica particles can act as reaction sites to anchor organic molecules. In Fig. 2(b), the band due to silanol groups entirely vanished, indicating that 3-APS molecules were bonded to the silica surfaces. While a new absorption band at 2120 cm^{-1} which can be attributed to azide groups (-N$_3$) appeared,[11,12] no stretching mode due to carboxyl group (-COOH) in 4-azidobenzoic acid was observed. These results confirmed that binding of the phenylazide groups (-C$_6$H$_4$-N$_3$) on the silica particles as shown in Scheme 1. When a phenylazide group is exposed to UV-light, it forms a nitrene group that can initiate inserting into N-H sites,[13] and the photoreactivity can be utilized for formation of covalent linkages among the particles.

The Az-silica particles were immobilized onto a glass substrate by photoactivation for evaluating their binding ability. Silane coupling agents chemisorb onto glass substrates resulting in a closely packed organic thin film in the form of a self assembled monolayer (SAM). A glass substrate was treated with 3-APS in accordance with a general SAM preparation protocol to

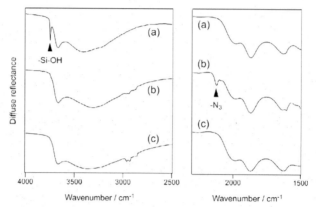

Figure 2. IR diffuse reflectance spectra of (a) silica particles as received, (b) Az-silica and (c) a green body after UV-light irradiation.

create a substrate coated with amino groups.[14] Az-silica particles were cast on the glass substrate coated with SAM and irradiated with UV-light (366 nm, 10.0 W) for 5 minutes. When the Az-silica particles are exposed to UV-light, they can be linked to neighboring amino groups by phenylhydrazo bondings (-C₆H₄-NHNH-). The reaction scheme for the immobilization of the Az-silica particles is shown in Scheme 2. The unreacted particles were removed by scotch tape separation. Figure 3 shows SEM images of the glass substrate surface after the scotch tape test.

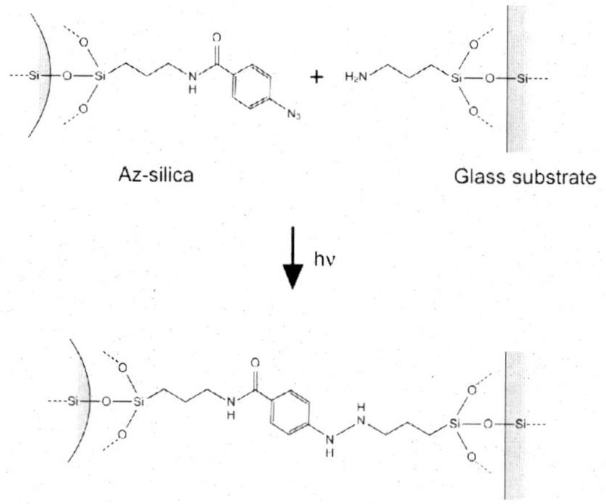

Az-silica   Glass substrate

hν

Scheme 2. Pathway for the covalent immobilization of the silica particles upon a glass substrate covered with amino groups.

Only the particles directly upon the substrate remained, as a monolayer, indicating that the immobilization reaction occurred between the AZ-silica particles and the glass substrate.

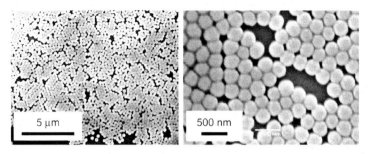

Figure 3. SEM images of the silica particles immobilized upon the glass substrate surfaces.

Mixture of equal parts of APS-silica and Az-silica (2.0 g) were dispersed in 20 ml of ethanol. The slurries were formed into tablets (diameter: 10 mm, height: 2 mm) by slip casting. The obtained tablet-shaped green bodies were exposed to UV-light (366 nm, 10.0 W) for 1 hour. The assumed covalent linkages among the silica particles are identical to that shown in Scheme 2. Figure 2(c) shows the IR spectrum of the obtained green body. Since the phenylazide groups were connected to neighboring amino groups by photoactivation, the absorption band of azide groups diminished when compared with that observed with Az-silica. This result confirmed that the covalent linkages due to photoactivation were successfully incorporated into the green bodies.

The obtained green body with covalent linkages and a green body as a control specimen (without any chemical bondings among the silica particles) were soaked in water and kept for a long duration. Any remaining air in the green bodies was expelled by keeping the whole assembly under reduced pressure atmosphere. When the green bodies are soaked in water, capillary condensation force among the particles due to adsorbed water should disappear.[15] In this case, green bodies without any other attractive force acting among their constituent particles than capillary condensation should not be able to maintain their shapes. While the control specimen was collapsed immediately after soaking in water, the green bodies with covalent linkages maintained their shape in water for up to 50 days (Fig. 4).

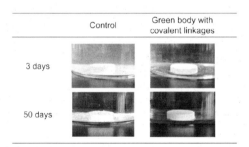

Figure 4. Photos of the green bodies soaked in water.

Figure 5 shows typical burnout profiles of the green body measured by thermogravimetry (TG) analysis. Weight increase of the control specimen above 400 °C should be due to oxidation of Si that remained imperfectly oxidized in the Si-O-Si silica network structure. Oxidation of the organic substances in green bodies starts at about 200 °C and weight loss below this temperature is due to dehydration.[16] The majority of the organic substances in the green body was removed between 200-400 °C. The shape of the green body were maintained in the presence of only 0.5 mass% organic substances, which was much smaller than usual polymer based molding systems. As long as conventional methods are employed, the amount of organic binders cannot be less than 1.0 mass%. Hence, it can be stated that the present method can reduce the amount of organic binder by half compared to that used in the conventional methods through linking of the particles under thin-film coverage conditions.

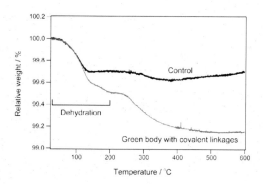

Figure 5.   TG curves of the green bodies.

The results mentioned above revealed that such a use of the organic binder as thin-film and the stronger bonding due to chemisorption result in significant reduction of the amount of organic binder required for retaining the shape of the green bodies. The present method is also applicable to other processes, for instance, immobilization of particles to prepare photonic or colloidal crystals. Selecting adequate coupling agents depending on the utilized ceramic phases is important to prevent the metal elements in the coupling agents from turning into impurities.

MICROWAVE-REACTIVE ORGANIC BINDER

For a widespread acceptance of this method, it is necessary to identify reaction triggers that can penetrate deeper into the ceramic green body structure than can UV-light irradiation. Because the penetration of UV-light into the ceramic materials is limited, it cannot be employed as a trigger for developing mutual linkages among the ceramic particles in bulk materials. As a replacement for UV-light, we investigated the use of microwave irradiation as a reaction trigger. Owing to its long wavelengths, microwaves can penetrate deep into the green bodies. However, the quanta of electromagnetic energy in the microwave spectral region are far too weak for direct excitation of electronic or vibrational transitions in molecules. Although the direct utilization of microwaves as a source of energy in chemical transformations has been reported,[17] its principle is still vague and it is accompanied by technical difficulties. In the present study, a macromolecule containing carbodiimide groups (–N=C=N–) and water-attracting (hydrophilic) segments in its structure was employed as a linking agent to form the particle assembly. The carbodiimide group can react with a variety of chemicals.[18-20] When green bodies containing the macromolecule within their structure are irradiated with microwaves, water molecules near the hydrophilic segments are dielectrically heated in a time-efficient manner. Subsequently, the increase in the internal temperature of the green bodies induces a reaction of the carbodiimide groups. By this method, we prepared green bodies composed of mutually connected ceramic particles by using microwaves as the reaction trigger.

2.5 g of APS-silica described in the precedent chapter was dispersed in ultra-pure water (25 mL) by using intense agitation. A water-dispersible polycarbodiimide (WDC; 2.5 g), from Nisshinbo Industries, Inc., Tokyo, Japan, was added to the suspension to link the APS-silica particles. On irradiation with the microwaves, the WDC forms covalent bonds with the amino-functionalized silica, binding the particles together: the carbodiimide group reacts with the amino group to form a guanidine structure when it is heated to ~80 °C (Fig. 6).[21] The WDC employed has reactive carbodiimide segments (–N=C=N–) and water-attracting (hydrophilic) oxyethylene segments (–CH$_2$CH$_2$O–): the molar ratio of carbodiimide to oxyethylene segments is approximately 1:10. The slurry was formed by slip casting into tablets of diameter 10 mm and height 2 mm and into rectangular 10 × 10 × 4 mm solids.

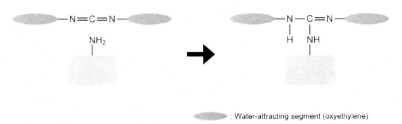

: Water-attracting segment (oxyethylene)

Figure 6.   Binding mechanism of a polycarbodiimide reactive segment with amino group upon microwave irradiation.

Microwave irradiation was carried out in a tunable magnetron multimode microwave furnace (MW-Master, Mino Ceramics Co., Ltd., Gifu, Japan).[22] The green bodies were placed in a thermally insulated box and irradiated with 200 Wh of microwaves at 2.45 GHz in the presence of microwave-absorbing dummy loads consisting of glass bottles, each filled with a mixture of poly(vinyl alcohol) flakes and alumina powder.

In IR spectrum of as-received WDC, a diagnostic band from carbodiimide stretching is found at 2120 cm^{-1}. IR diffuse reflectance spectra of the microwave-irradiated green bodies in the carbodiimide stretching region are shown in Fig. 7. When carbodiimide groups are heated to ~80 °C, they initiate bonding to neighboring amino groups on the silica surface, and this reactivity can be used to form covalent linkages between the particles. Decreases in the strength of the carbodiimide absorption bands can be used to monitor the progress of the reaction: the intensity of the absorption band decreased with increasing time of microwave irradiation. The decrease in the intensity of the absorption band indicated the formation of linkages between the silica particles.

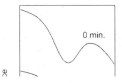

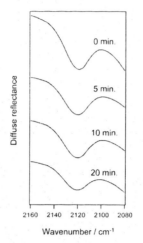

Figure 7. Spectral changes in the carbodiimide absorption band during microwave irradiation. The elapsed times are 0, 5, 10 and 20 min, from top to bottom.

We prepared two kinds of green bodies, one with and one without covalent linkages. The former was a green body containing WDC treated by microwave irradiation for 20 min, and the latter was a green body prepared in the same way, but using poly(ethylene glycol) (PEG) instead of WDC. PEG is composed solely of oxyethylene segments and contains no carbodiimide segments. Because there is no chemical bonding between the silica particles, the PEG-containing green body can be used as a control specimen. The two kinds of specimen were soaked in water and kept for a long period. Remaining air in the green bodies was expelled by keeping the whole assembly under a reduced-pressure atmosphere. Green bodies with no attractive force other than capillary condensation acting between the constituent particles

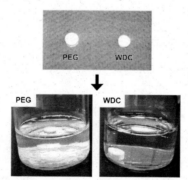

Figure 8. Photographs of the green bodies prepared by using PEG and WDC. The green bodies were soaked in water to evaluate the advantageous effect of interparticle bonds induced by microwave irradiation.

should not be able to maintain their shapes. Whereas the control specimen collapsed immediately on soaking in water, the green bodies with covalent linkages maintained their shape in water (Fig. 8).

The mechanical properties of the green bodies were evaluated by using a universal testing machine. Because the green bodies were not subjected to any firing process, the usual mechanical evaluation methods for ceramic sintered bodies were not applicable. Hence, in our study, we employed a simplified breaking test. The total weight on the load cell was recorded when the push-pin broke the rectangular green bodies. We tested 18 specimens of green bodies containing WDC and a similar number of control specimens and recorded the weight values when the green bodies were broken. The results are summarized in Fig. 9. The mechanical strength of the control specimen was, naturally, poor. Disintegration began at a load of about 500 g. Disintegration of the green bodies with the chemical bonds occurred at loads of more than 1000 g, and the mechanical properties of these green bodies were significantly better than those of the control specimens.

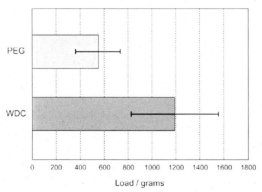

Figure 9. Evaluation of the mechanical properties of the green bodies.

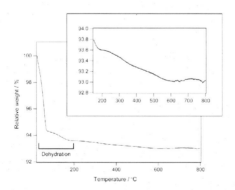

Figure 10. TG curve of the WDC-containing green body. The region relating to the burning out of organic substances is magnified.

Figure 10 shows typical burnout profiles of the WDC-containing green body, measured by TG analysis. Because the oxyethylene segments in WDC are highly hydrophilic, the green body still contained significant amounts of water even after 20 min of microwave irradiation. The majority of the organic substances in the green body were removed between 200 and 600 °C. The shapes of the green bodies were maintained in the presence of only 0.6 mass% of organic substances, which is much less than required for conventional polymer-based molding systems. With such a small amount of binder, green bodies are usually too fragile to handle. Therefore, the present method, through linking of the particles induced by microwave irradiation, results in a marked reduction in the amount of organic binder that is required compared with that used in conventional methods.

SUMMARY

A novel ceramic-forming process was proposed in this article. Photoreactive and microwave-reactive organic molecules were used to form bridges to bind the entire particle assembly. The resultant assembly can be considered as a ceramic green body. This unique methodology results in a significant reduction in the quantity of organic binder that is required to maintain shape, while providing a stronger binding as a result of the chemical forces that are involved. The green bodies obtained in this study displayed improved mechanical properties in the presence of only minimal amount of organic substances. The present method should surely contribute to a reduction of organic binders required for green body preparation when compared to the currently utilized conventional methods.

ACKNOWLEDGMENT

The present work has proceeded as a part of the collaborative research project entitled "Research and Development of Low Environmental Load Processes" between AIST and NGK Insulators Ltd., Japan.

REFERENCES

[1]M. Bengisu, Engineering Ceramics; pp. 85-207. Springer, Berlin, Germany, 2001.

[2]W. E. Lee and W. M. Rainforth, Ceramic Microstructures: Property Control by Processing; pp. 3-66. Chapman & Hall, London, UK, 1994.

[3]F. F. Lange, "Powder Processing Science and Technology for Increased Reliability," *J. Am. Ceram. Soc.*, **72**, 3-15 (1989).

[4]K. Prabhakaran, S. Ananthakumar, C. Pavithran, "Gel Casting of Alumina Using Boehmite as a Binder," *J. Eur. Ceram. Soc.*, **19**, 2875-2881 (1999).

[5]T. Nagaoka, T. Tsugoshi, Y. Hotta, K. Watari, "Fabrication of Porous Alumina-based Ceramics with Hydraulic Inorganic Binder," *J. Ceram. Soc. Japan, Suppl.*, **112**, S277-S279 (2004).

[6]T. Nagaoka, T. Tsugoshi, Y. Hotta, K. Sato, K. Watari, "Fabrication of Porous Alumina Ceramics by New Eco-Friendly Process," *J. Ceram. Soc. Japan*, **113**, 87-91 (2005).

[7]T. Nagaoka, T. Tsugoshi, Y. Hotta, M. Yasuoka, K. Watari, "Effects of Alumina Hydrates Formed by Hydration of Hydraulic Alumina on Green Strength and Microstructure of Porous Alumina Ceramics," *J. Ceram. Soc. Japan*, **114**, 214-216 (2006).

[8]B. Arkles, J. R. Steinmetz, J. Zazyczny, P. Mehta, "Factors Contributing to the Stability of Alkoxysilanes in Aqueous Solution," *J. Adhesion Sci. Technol.*, **6**, 193-206 (1992).

[9]C. R. Suri, G. C. Mishra, "Activating Piezoelectric Crystal Surface by Silanization for Microgravimetric Immunobiosensor Application," *Biosens. Bioelectron.*, **11**, 1199-1205 (1996).

[10]M. Nisnevitch, M. Kolog-Gulco, D. Trombka, B. S. Green, M. A. Firer, "Immobilization of Antibodies onto Glass Wool," *J. Chromatogr. B*, **738**, 217-223 (2000).

[11]A. Zhu, M. Zhang, J. Wu, J. Shen, "Covalent Immobilization of Chitosan/Heparin Complex with a Photosensitive Hetero-bifunctional Crosslinking Reagent on PLA Surface," *Biomaterials*, **23**, 4657-4665 (2002).

[12]C. Mao, A. P. Zhu, Y. Z. Qiu, J. Shen, S. C. Lin, "Introduction of *O*-butyrylchitosan with a Photosensitive Hetero-bifunctional Crosslinking Reagent to Silicone Rubber Film by Radiation Grafting and Its Blood Compatibility," *Colloid Surface* B, **30**, 299-306 (2003).

[13]G. T. Hermanson, Bioconjugate Techniques; pp. 252-255. Academic Press, San Diego, CA, 1996.

[14]J. Sagiv, J. *Am. Chem. Soc.*, "Organized Monolayers by Adsorption. 1. Formation and Structure of Oleophobic Mixed Monolayers on Solid Surfaces," **102**, 92-98 (1980).

[15]J. Israelachvili, Intermolecular & Surface Forces, second edition; pp. 330-334. Academic Press, London, UK, 1992.

[16]A. C. Young, O. O. Omatete, M. A. Janney, P. A. Menchhofer, "Gelcasting of Alumina," *J. Am. Ceram. Soc.*, **74**, 612-618 (1991).

[17]R. Maoz, H. Cohen, J. Sagiv, "Specific Nonthermal Chemical Structural Transformation Induced by Microwaves in a Single Amphiphilic Bilayer Self-assembled on Silicon," *Langmuir*, **14**, 5988-5993 (1998).

[18]M. Mikołajczyk, P. Kiełbasi´nski, "Recent Developments in the Carbodiimide Chemistry," *Tetrahedron*, **37**, 233-284 (1981).

[19]A. Williams, I. T. Ibrahim, "Carbodiimide Chemistry: Recent Advances," *Chem. Rev.*, **81**, 589-636 (1981).

[20]L. C. J. Hesselmans, A. J. Derksen, J. A. M. van den Goorbergh, "Polycarbodiimide Crosslinkers," *Prog. Org. Coatings*, **55**, 142-148 (2006).

[21]EU Patent 0878496, JP Patent 10316930, US Patent 6124398, Y. Imashiro, I. Takahashi, N. Horie, T. Yamane, S. Suzuki, to Nisshinbo Industries Inc., 15 May 1998.

[22]M. Yasuoka, T. Shirai, Y. Nishimura, Y. Kinemuchi, K. Watari, "Influence of Microwave Irradiation Method on the Sintering of Barium Titanate with Liquid Phase," *J. Ceram. Soc. Japan*, **114**, 377-379 (2006).

DRYING DINETICS OF SLIP CAST BODY BY MICROWAVE HEATING

Takashi Shirai, Masaki Yasuoka, Yoshiaki Kinemuchi, Yuji Hotta, and Koji Watari
National Institute of Advanced Industrial Science and Technology (AIST),
2266-98 Anagahora, Shimoshidami, Moriyama, Nagoya 463-8560, Japan

ABSTRACT

We investigate that a drying kinetics of slip cast body in a ZnO slurry between microwave heating and conventional heating. The results of the present study showed that the drying stage could be divided into two stages in microwave heating. Microwave heating can selectively heat the water in a slurry, and a small addition of NH_4^+ salt of poly acrylic acid (PAA) to the water can have important effects on microwave absorption. Microwave heating can greatly improve the drying kinetics due to selective heating, and the drying stage can be reduced to two steps. The transport of water during drying was enhanced in the linear kinetics stage through the reduction of water surface tension due to selective heating by microwaves.

INTRODUCTION

The slip casting method [1-4] is an essential technique in producing technical and/or advanced ceramics. A major disadvantage of slip casting is that it requires long periods for forming a body of the desired thickness and further periods to dry the cast body. Conventional drying of ceramics usually will take a long time. Thus, there have been many attempts to increase the rate of drying of ceramics in order to decrease drying time.

The objective of drying is simply to remove water from the ceramic body without causing any damage. The process must be done both efficiently and economically. Water can leave the surface of the ceramic at a given rate depending on relative humidity and temperature. To accomplish good drying requires a method that removes the water from the inside of the ceramics to the outside surface at the same rate as the evaporation of surface water. Therefore, the drying must be conducted either in an extremely slow rate so as to preserve the piece intact [5]. In addition, it is effective to use a drying control chemical and/or with a nonaqueous solvent to enhance diffusion of water [6].

A practical way to distinguish between the various drying processes is to classify them by the heating mode; convective, radiative, and conductive. Drying with internal heat generation, such as dielectric drying, is a special case. We believe that one of the most important fields of application of microwave technology is drying. The high loss factor of water, the volumetric energy absorption and unlike the conventional drying the inverse thermal gradient and because the most wet parts of the material are at the same time the most hot ones, provide favourable conditions for drying.

The objective of this study is to demonstrate the applicability of microwave drying for rapid drying technique in a slip cast body. The study was focused on the differences in the water evaporation behaviour of a ZnO slurry using microwave drying and conventional drying.

EXPERIMENT

A commercially available sub-micron ZnO powder (JIS 1, Hakusuitech Co., Ltd.) with an average diameter of $d_{50}=0.5$ μm was used for the preparation of slurries that contained 40 vol% solid phase and 0.32 mass% of a commercially available NH_4^+ salt of poly acrylic acid (PAA, Aron A-6114, Toagosei Co., Ltd.) as a dispersant.

After 2 min of mixing with an electric mixer (AR-250, THINKY Corp.), the suspension was ball-milled for 12 h with high-grade Al_2O_3 balls (diameter 5 mm) to prepare the ZnO slurry.

The slip was cast on a gypsum mold to form green bodies ($20\times70\times t3$). Before their removal from the mold, they were dried by various drying methods. In oven drying, a green body was dried at $80\pm3°C$ (RH 40%) in an oven for at least 24 h. This drying period appeared to be sufficient to evaporate all free water and most bound water.

In air-drying, a green body was dried at room temperature ($25\pm3°C$, RH 40%) in an incubator for at least 48 h. Air-drying required a longer period to reduce water content to an almost constant value. A microwave oven (MW-Master, 2.45 GHz, Mino Ceramic Co. LTD.,) was used for microwave drying. Drying temperature was controlled by a proportional integral differential (PID) controller. A green body was heated to $80\pm3°C$ at a constant heating rate of 2 K/min, and kept at this temperature for 30 min. All free water and most bound water evaporated rapidly in the microwave furnace. After drying, the green bodies were sintered by a conventional method. The sintering temperature was $1000°C$, the heating rate was 10 K/min and the holding time was 2 h.

The densities of green and sintered bodies were measured by the Archimedes method. The warpage of green bodies was evaluated using a laser confocal displacement meter (LT-9010M, Keyence Corp.). The microstructures of these bodies were observed on polished surfaces by scanning electron microscopy (SEM, JSM-5600N, JEOL Ltd.,).

RESULTS

Figure 1 shows the extent of warpage of the dried bodies examined on their bottom face using the laser confocal displacement meter. The extent of warpage of the dried body is smaller when dried by microwave than by the other conventional drying methods.

Figure 2 shows a SEM image of the polished surface of the sintered bodies. The bottom part of the air-dried sample, Fig. 2-(a), shows many large pores. Local differences in water content arise during the long drying period of the air-dried green body, forming many pores on the bottom of the part. In oven drying, many large pores occurred in the upper part of the green body particularly at the surface as seen in Fig. 2-(b). In this case, the green body is heated from the outside generating a temperature difference in the green body, resulting in rapid evaporation of free water occurs from the upper surface, producing many large pores. With microwave drying, the pores are smaller in number and size than those obtained with other drying methods, moreover, the pores were distributed uniformly.

Figure 3 shows a mass loss with drying time for the ZnO slurry for the different drying techniques. In conventional drying, Fig. 3-a, a minimum of 9 h is required to bring the water content to an almost constant value. The remaining water could be mostly ascribed to the water at the meniscus regions and adsorbed water. On the other hand, microwave drying, Fig. 3-b, can eliminate the water to the same limits in only 1.6 h.

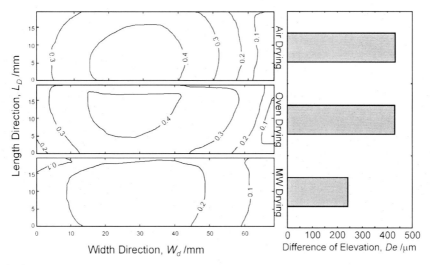

Figure 1: Extent of warpage of dried bodies dried by different drying methods.

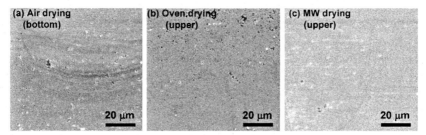

Figure 2: SEM image of polished fractured surface after sintering; (a) air drying in bottom part, (b) oven drying in upper part, and (c) microwave drying in upper part, respectively.

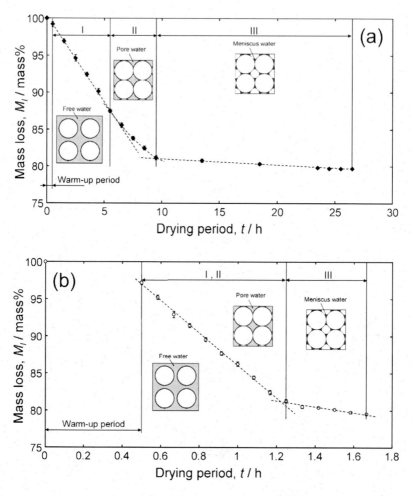

Figure 3: Comparison of conventional drying (a) and microwave drying (b) on the water evaporation behavior. In conventional drying, the weight loss curve can be divided into three steps. In comparison, by microwave drying, the weight loss curve can be separated into only two steps.

DISCUSSION

Drying of green body has recently become more important due to the use of ultrafine powder. The capillary size of the porosity in the green body reduces considerably as the particle size of the powder decreases. The transport of water through the capillaries less than 1/5 of the particle diameter exerts a strong drying stress on the green body. Therefore, the drying must be conducted either at an extremely slow rate so as to preserve the body intact. The evaporation and transport of water are the keys to drying and controlling the kinetics of drying, subsequently affecting the properties of green body [13].

In conventional drying, shown in Fig. 3-a, the drying rate was fixed at first, and it decreases gradually with drying time. The mass loss curve can be separated into three steps. Generally speaking, in the case of a slip cast body, the drying process by conventional heating follows three steps [5]. The green body consists of particles and water before drying. In the first step, the free water around the particles (as shown of the diagram A in Fig. 3-a) evaporates, and the resulting shrinkage is due to the decrease in the distance between particles. After that, the drying process shifts to the second step in which particles make contact with each other. The turn point of the curves to the second step of drying in Fig. 3 is close to a mass loss of ca. 12 mass% that is corresponded to green density after drying. In the second step, the pore water (as shown of the diagram B in Fig. 3-a) is transported to the surface reducing the total content in the green body. However, a water film will still remain in the contact region (as shown of the diagram C in Fig. 3-a) forming a meniscus shape. Third step starts as the evaporation of the water at the meniscus regions becomes dominant [5].

Drying is basically an interface control process. The drying rate is theoretically controlled by the evaporation at the outer surface. As long as the environmental conditions and the surface area of the gas/water interface remain constant, the drying rate is fixed [13]. Therefore, the first step has a constant drying rate which means the rate of water loss is constant with respect to time. However, in the second step, the drying rate is decreased by reduction of the capillary size due to shrinkage. A parabolic relation between the drying time and mass loss is observed which means this step is diffusion controlled drying, as shown in the second step of Fig. 3-a. The evaporation and transport of water are the keys to drying and controlling the kinetics of drying, subsequently affecting the properties of green body [13]. Therefore, drying must be conducted at an extremely slow rate for the body to remain intact.

In comparison, for the case of microwave heating, the drying proceeds in two steps, Fig. 3-b. The mass loss curve for microwave heating has no distinct boundary between first step and second step. These steps have a constant drying rate which means the rate of water loss is constant with respect to time.

Microwave heating is fundamentally different from conventional heating. In microwave heating, heat is generated internally within the material instead of from external heating sources [7]. A material needs to have a high dielectric loss to absorb microwaves efficiently. Figure 4, reprinted from our previous study [14], shows the temperature-time curves for the ZnO slurry, water with PAA, water, and a ZnO powder under the following conditions: f=2.45 GHz, 150 W. After MW irradiation for 6 min, the ZnO slurry and the water with PAA reached 75°C and 71.4°C, respectively. The results for the ZnO slurry and water with PAA showed similar temperature-time curves. On the other hand, the temperature of the ZnO powder only slightly increased to 27°C. From these results, the ZnO powder did not undergo an increase in temperature under microwave irradiation. In the case of the ZnO slurry, the water with PAA was heated selectively due to a higher dielectric constant compared with the ZnO powder.

Comparing the water with PAA to only water, the rate of temperature increase for the water with PAA was higher than that of only water, and reached a temperature almost 20°C higher than that reached by only water after 6 min. It is believed that a small addition of PAA to water has important effects on microwave absorption. From these results, the water with PAA in slurry is heated selectively by microwaves. This means that the temperature distribution of green body is small due to selective heating by microwave, because it can be considered that water is distributed in a green body equally. Besides, in the case of conventional heating, the green body has lower internal temperature than outside temperature, because green body is heated from the outside. On the other hand, in the case of microwave heating, the green body has higher internal temperature than outside temperature, because green body is heated by itself due to selective heating. The disappearance of the diffusion control process by microwave heating in Fig. 3-b is the transport of the water during drying, enhanced in the stage of linear kinetics through the reduction of water surface tension due to selective heating by microwave.

Based on these results, microwave heating can greatly improve the drying kinetics due to selective heating and the drying stage can be reduced to two steps. Therefore, the drying defects such as warpage, cracks and density distribution are fewer by microwave drying than by conventional drying, and the microstructure of the sintered bodies are more uniform when dried by microwaves, as reported in our previous results [12].

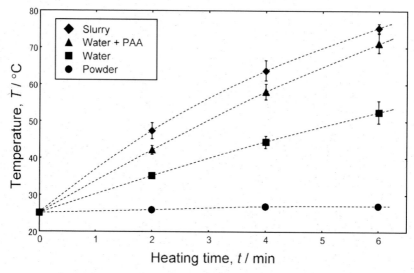

Figure 4: Temperature-time curves for the ZnO slurry, water with PAA, water, and a ZnO powder under the following condions: f=2.45 GHz, 150 W [14].

CONCLUSIONS

We observed a difference in the water evaporation behaviour in a ZnO slurry between microwave heating and conventional heating. The following conclusions were drawn.

(1) Water with PAA in the slurry is heated selectively with microwave drying. A small addition of PAA to water has important effects on microwave absorption.

(2) Microwave heating can greatly improve the drying kinetics due to selective heating and the drying stages can be reduced to two steps.

(3) The transport of the water during drying is enhanced in the linear kinetic stage through the reduction of water surface tension due to selective microwave heating.

(4) Microwave heating results in a small gradient of the water content between the inside and the outside of a green body, because the evaporation of water is more efficient. Therefore, the extent of warpage of the dried body is smaller by microwave drying than by conventional drying, and the resulting microstructure of the sintered bodies is uniform.

REFERENCES

[1]Y. Hotta, T. Tsugoshi, T. Nagaoka, M. Yasuoka, K. Nakamura, and K. Watari, "Effect of Oligosaccharide Alcohol Addition to Alumina Slurry and Translucent Alumina Produced by Slip Casting", *J. Am. Ceram. Soc.*, **86**, 755-760 (2003).

[2]K. S. Chou, and L. J. Lee, "Effect of Dispersants on the Rheological Properties and Slip Casting of Concentrated Alumina Slurry", *J. Am. Ceram. Soc.*, **72**, 1622-1627 (1989).

[3]L. B. Garrido, and E. F. Aglietti, "Pressure Filtration and Slip Casting of Mixed Alumina-Zircon Suspensions", *J. Eur. Ceram. Soc.*, **21**, 2259-2266 (2001).

[4]N. Omura, Y. Hotta, K. Sato, Y. Kinemuchi, S. Kume, and K. Watari, "Slip Casting of Al_2O_3 Slurries Prepared by Wet Jet Milling", *J. Ceram. Soc. Japan*, **113**, 495-497 (2005).

[5]Y. Shiraki, "Ceramic-Seizo-Process III", *Gihodo Shuppan*, pp. 39-43 (1980).

[6]I. W. Turner and P. G. Jolly, "Combined Microwave and Convective Drying of a Porous Material", *Drying Technol.*, **9**, 1209-1269 (1991).

[7]W. H. Sutton, "Microwave Processing of Ceramic Materials", *Am. Ceram. Soc. Bull.*, **68**, 376-386 (1989).

[8]D. Skansi, and S. Tomas, "Microwave Drying Kinetics of a Clay-Plate", *Ceram. Int.*, **21**, 207-211 (1995).

[9]R. Y. Ofoli, and V. Komolprasert, "On the Thermal Modeling of Foods in Electromagnetic Fields", *J. Food Processing and Preservation*, **12**, 219-241 (1988).

[10]P. Jolly, and I. W. Turner, "Non-Linear Field Solutions of One-Dimensional Microwave Heating", *J. Microwave Power and Electromagnetic Energy*, **25**, 3-15 (1990).

[11]P. Jolly, "Temperature Controlled Combined Microwave- Convective Drying", *J. Microwave Powder*, **23**, 65-74 (1986).

[12]T. Shirai, M.Yasuoka, Y. Hotta and K. Watari, "Rapid Microwave Drying for Slip Cast Bodies", *J. Ceram. Soc. Japan*, **114**, 217-219 (2006).

[13]Wen-Cheng J. Wei and Chang-Li Hsieh, "Drying Kinetics of Ultrafine Alumina Cake with Drying Control Cemical (DCC)", *J. Ceram. Soc. Japan*, **107**, 313-317 (1999).

[14]T. Shirai, M. Yasuoka, Y. Hotta, Y. Kinemuchi and K. Watari, "Microwave Drying for Slip Cast Body", *Adv. In Tech. of Mat. Proc. J. (ATM)*, (2006), in press.

MICROWAVE SINTERING TECHNIQUES - MORE THAN JUST A DIFFERENT WAY OF HEATING?

G. Link, S. Miksch, M. Thumm*
Forschungszentrum Karlsruhe, IHM
Hermann-von-Helmholtz-Platz 1,
76344 Eggenstein-Leopoldshafen, Germany
*and University of Karlsruhe , IHE, Germany

S. Takayama
National Institute for Fusion Science
322-6 Oroshi-Cho Toki
Gifu, 509-5292 Japan

ABSTRACT

The use of microwaves for sintering has been proposed and investigated by several research groups, because it allows a direct energy transfer into the material's volume and therefore allows an essential acceleration of the heating process. Furthermore several publications in this field conclude an ʻenhanced densification from a possible reduction of sintering temperatures and/or soak time if microwaves are used for heating instead of gas fired or resistance heated furnaces. Such phenomenological observations are usually explained by so called non-thermal microwave effects. But quite often possible errors in temperature measurement resulting from intrinsically different temperature gradients were not taken into account, when microwave and conventional heating are compared.

Therefore a novel experimental approach has been developed which allows a direct experimental access to non-thermal microwave effects. Based on the theory of the so called ponderomotive driving forces which specifies an enhanced diffusion in ionic solids under the influence microwave fields, the influence of the microwave field orientation onto the diffusion in a faceted pore has been described by Booske et al. [1]. In consequence of this, an anisotropic pore closure can be expected during sintering of ceramics in a linearly polarized microwave field. Systematic investigations of the pore structure evolution in yttria stabilized zirconia have been started in a single mode 2.45 GHz waveguide applicator. For the first time, strong experimental evidence for the existence of an anisotropic pore closure due to a non-thermal microwave effect was found with an adequate statistic evaluation of the pore aspect ratios after sintering.

INTRODUCTION

One of the major disadvantages of conventional sintering (CS) in standard resistance heated or gas fired furnaces is that temperature gradients are induced into the material due to the low penetration depth of IR-radiation. Therefore conventional heating has to rely on the thermal conductivity of the material, which is the key factor that determines how fast and homogeneous materials can be heated. If not an optimized time-temperature program with sufficiently low heating rate is applied for sintering of ceramics, temperature gradients may evolve leading to thermal stresses, differential shrinkage or crack formation. The use of microwaves allows a direct energy transfer into the material's volume, where it is converted to heat through absorption mechanisms such as ionic conduction, dipole relaxation and photon-phonon interaction. Here the key factors for fast and homogeneous heating are the distribution of the electromagnetic field within the applicator and the material, respectively as well as the microwave absorptivity of the material. Thermal conductivity for the microwave heating process is secondary.

Worldwide the heating and sintering process of ceramics has been studied comprehensively for a large variety of ceramics using different microwave frequencies and various microwave furnaces. Beside the procedural benefit of instantaneous volumetric heating enhanced densification has been found during sintering in microwave fields [2-8]. This usually manifests as the potential for lowering the sintering temperature and soak time compared to conventional processing in a resistant heated of gas fired sintering furnace. But very often there is no information given about existing temperature gradients or systematic errors in temperature measurement. Very easily large errors can be disregarded due to the fact that temperature sensors, which are used in the compared sintering systems, are not calibrated. Or even bigger errors may happen due to the fact that temperature gradients evolve within the ceramic sample during heating which in standard experiments are not accessible.

In order to demonstrate these sources of errors in temperature measurement recently a system allowing detailed comparative investigations of different heating methods has been utilized [9]. This is a compact 30 GHz gyrotron system equipped with a modular heating system. which allows conventional sintering (CS) by resistive heating, millimeter-wave sintering (MWS) as well as millimeter-wave assisted heating, so called hybrid sintering (HS) up to a temperature of 1650°C. And this module has been used in combination with a dilatometer and a multi-channel temperature monitoring system using several type S shielded thermocouples, so that the evolution of temperature gradients can be monitored during the heating process [10]. For the following experiments three thermocouples were installed, one monitored the temperature T_{wall} near by the $MoSi_2$ heating element in the wall controlled by conventional heating, another monitored the temperature T_{out} measured at the sample surface and the last monitored the temperature T_{in} inside the sample through a small hole drilled into the sample before sintering (see Fig. 1). The data acquisition of the last two thermocouples provides information about the evolution of temperature gradients within the sample during the heating process.

All these heating possibilities have been used to measure dilatometer curves from ceramic powder compacts made from a submicron zirconia powder of the type TZ3Y-SE from the Tosoh Company, Japan. The size of the cylindrically shaped compacts was 6.4 mm in diameter and about 10 mm in length. The results of dilatometer experiments with a heating rate of 20 K/min. up to the sintering temperature of 1600 °C and 10 min. isothermal soak are shown in Figure 2. The left graph gives the measured linear shrinkage for all different heating methods as a function of that temperature typically used to control the process. So results of CS are plotted as a function of the temperature T_{wall} and results of MWS as well as HS are plotted as function of the temperature T_{out} measured at the sample surface. Therewith sintering with millimeter waves obviously starts at temperatures 300 °C lower than sintering with conventional heating. The dilatometer curve obtained during hybrid sintering lies in-between conventional and millimeter wave sintering. If we plot the same results by using T_{in} as the reference temperature (see right graph of Figure 2), that was measured in the sample volume, then the curve for HS stays more or less at the same position and the dilatometer curves obtained by MWS and CS, respectively are much closer. This clearly reveals the problem of temperature gradients. From these experiments it is not clear if the residual difference between CS and MWS is due to an enhanced sintering in the microwave field or just due to the fact that the location where the temperature was measured was not representative for the process.

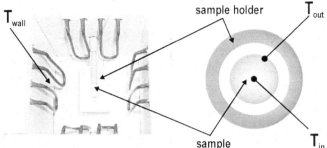

Fig. 1. View into the hybrid heated ceramic fiber box including the dilatometer (left) and schematic cross-section of the sample holder indicating the location of temperature measurement (right). [9]

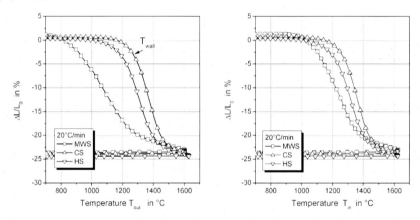

Fig. 2. Dilatometer curves for different heating methods with T_{out} as temperature axis (left) and T_{in} as temperature axis (right) [9]

For further clarification of any microwave specific effects all sample microstructures were characterized by SEM on polished and thermally etched crosscut samples surfaces. The average grain size was estimated. Within the accuracy of this method, no significant difference in grain size could be found, neither along the cross-section of a single sample nor for the different ways of heating. Since the different heating methods usually result in inherent different heating profiles, the usually measured temperature clearly underestimates the effective temperature in case of microwave sintering and overestimates the effective temperature in case of conventional heating. If such sources of errors are not carefully taken into consideration, the argumentation for non-thermal microwave effects or microwave-enhanced diffusion might be insincerely. Therefore this topic is very often a matter of criticism [11].

However a successful commercialization of such a novel technology must be based on an improved and reliable fundamental knowledge of microwave specific mechanisms resulting in such promising phenomenological observations. That's why recently more specific experimental results on microwave processing were published, related to microwave induced mass transport in ionic solids which hardly can be explained by thermal driving forces [12-15]. Some of these experiments were also found to be in good agreement with existing theoretical models [16, 17]. This most commonly used theory is based on a model originating from plasma physics which describes a net time-averaged mass flow of charged plasma particles moving in an oscillating inhomogeneous electromagnetic field. The resulting driving force is named ponderomotive force. This model has been adapted to ionic materials exposed to microwaves, where the diffusion or mass transport in solids is based on a non-zero time-averaged ponderomotive driving force to ions moving in an oscillating inhomogeneous concentration of such ionic species, induced by the microwave fields especially at grain boundaries [18].

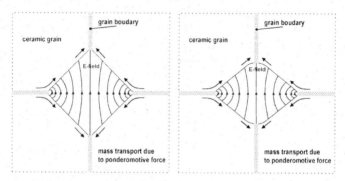

Fig. 3. Scheme of a ceramic pore indicating field distribution and non-thermal mass transport according to Booske et al. [1] at an initial (left) and medium (right) stage of sintering.

As a further refinement of these theoretical models, the effect of polarized electric fields onto the mass transport in ceramics has been analyzed in the vicinity of faceted pores [1]. In combination with a field enhancement in pores, as it has been derived by Birnboim et al. [19] in spherical neck regions in ceramics prior to full densification, an anisotropic ion flux and therefore an anisotropic mass transport can be predicted. As a consequence of this microwave sintering of ceramics in a linearly polarized microwave field should cause an anisotropic pore closure as schematically indicated in Figure 3. If such an effect really exists and if it is sufficiently large to be verified by experiments, this would be a unique method to demonstrate non-thermal microwave effects. Such an experimental proof will not have to rely on comparison with conventional heating and therefore there will be no need to rely on temperature measurement. In the remaining part of this paper, efforts on experimental verification of such an effect are described.

EXPERMENTAL

Experimental configuration

A standard TE$_{103}$ waveguide applicator is the simplest type of microwave applicator to investigate effects of polarized electric fields onto the pore structure evolution in ceramic materials. The utilized waveguide had a square cross section of 8 cm x 8 cm, which allows aligning the electric field of the TE$_{103}$ waveguide mode in horizontal as well as in vertical direction [20]. At both sides of the applicator a water-cooled 1.2 kW magnetron with a frequency of 2.45 GHz was installed in combination with a circulator and a three-stub tuner. The microwave power was launched via a standard WR340 waveguide and an adequate taper, with horizontal polarization of the electric field from one side and vertical polarization from the other side. The experiments described in the following were performed with vertical field polarization only, as it is indicated in Figure 4. By adequate positioning of a sliding polarizer on one side of the applicator and impedance matching by use of the three-stub tuner on the other side the maximum electric field of the TE$_{103}$ mode was moved to the sample position. To avoid temperature gradients during the sintering process, which may result in inhomogeneous sintering as well, the specimen were placed into a zirconia crucible, according to the thermal barrier wall concept of Sato et al. [21]. The crucible itself was surrounded by mullite ceramic fiberboards for thermal insulation. The lid of the crucible as well as the fiber board was provided with a 7 mm bore hole in order to enable a direct view of the pyrometer to the sample surface. The pyrometer signal was taken to control the sintering process using a constant heating rate of typically 30 K/min to the final sintering temperature. An appropriate soak time was applied so that the densities reached were in the range of 95 %TD to 99 %TD. This allowed having a microstructure with closed porosity and clearly separated pores for the subsequent microstructure analysis.

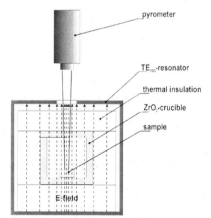

Fig. 4. Scheme of the experimental setup in the waveguide resonator.

Sample preparation

Since the choice of material was expected to be essential for the outcome of the intended investigations and since theoretical models so far are based on ionic materials, zirconia ceramic was chosen for testing. This material is well known for its high ionic conductivity especially at high temperatures as they are necessary for sintering. So a maximum of interaction with microwave fields can be expected. Yttria stabilized zirconia from Tosoh Company, Japan was used for sample preparation. It was a tetragonal zirconia, stabilized with 3 mol% yttria (TZ3YS) with a theoretical density of 6.07 g/cm^3. In order to be able to diminish the influence of superimposed thermal effects the tetragonal zirconia used was a nanosized powder with a mean grain size of 27 nm. Due to the higher sintering activity of nanosized powders compared to its coarse grained counterparts, reduced sintering temperatures can be applied and thus thermally activated driving forces might be smaller.

Another effect which might influence the evolution of the pore geometry and therefore the outcome of the proposed investigations is the powder handling, in particular the powder compaction technique. As it is known from constrained sintering of ceramic films or pressure assisted densification of bulk ceramics, sample preparation may result in anisotropic sintering which manifests in elongated pores and grains [22, 23]. Therefore to eliminate potential anisotropic strains or stresses within the green bodies, samples were not fabricated by uni-axial pressing but by the electrophoretic deposition technique (EPD) [24] at the University of Saarbrücken, Germany. For an optimal shaping with EPD a dispersion of the powders into an agglomerate-free suspension with a high solid loading was realized. The so produced green bodies were dried for 24 h at room temperature and then in a drying chamber at 120 °C for another 24 h. The samples had a green density of approximately 53 % of the theoretical density (TD). Hg-porosimetry revealed a homogenous and monomodal pore size distribution with a mean pore size of 45 nm for TZ3YS.

Sample characterization

After sintering, first of all the density of the samples was measured using the Archimedes Principle. Samples with adequate density for the microstructure analysis were then cut and polished. After thermal etching at 1100 °C for about one hour in a lab kiln a number of micrographs showing isolated pores were sampled, using a scanning electron microscope (SEM). The samples were oriented within the SEM in that way, that the electric field as present during microwave sintering could be indicated in vertical direction. Then all SEM snapshots were analysed with ImageJ free software (National Institutes of Health, USA) the following way. Before image analysis all micrographs were transformed in 8-bit greyscale images and appropriate threshold values were defined in order to obtain suitable binary images which separate all pores from grains and grain boundaries (see Figure 5). These binary images were then analysed in two different ways. One method fitted bounding rectangles. Therewith the maximum pore extensions perpendicular $d_{\perp E}$ and parallel $d_{\parallel E}$ to the orientation of the electric field could be identified. Then the aspect ratio $d_{\perp E}/d_{\parallel E}$ of each pore in correlation to the orientation of the electric field was calculated. For statistical analysis a log transform of the data was used to provide a normal distribution. In order to get reasonable statistical significance only samples were taken into further consideration if a minimum of about 50 different pores could be isolated. Finally three to four sampling procedures were used to derive a final mean value of the pore aspect ratio including a standard deviation.

A second method used the binary images to fit an equivalent ellipse to each pore. Then the angle between the large axis and the horizontal line, perpendicular to the direction of the electric field, was estimated. This information was then plotted in histograms showing the number of pores found in different intervals of angle. To reach sufficient statistical significance this method was applied only if the number of detectable pores was larger than 200.

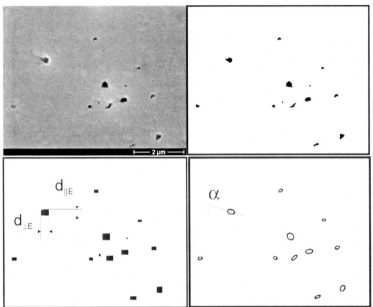

Fig. 5. ImageJ procedure: SEM picture (top left), conversions into binary image through definition of threshold value (top right), approximation by bounding rectangles (bottom left), approximation by ellipses (bottom right).

RESULTS AND DISCUSSION
Pore geometry in microwave sintered samples

For the first experiments the process parameters chosen followed the idea to reduce thermal effects in competition with non-thermal effects as much as possible. Therefore sintering temperatures were initially reduced to levels, where with still acceptable soak time reasonable densities could be achieved. So samples from TZ3YS were sintered at 1250 °C for seven days and at 1260 °C for 4 days, respectively. Thus sintered densities of 98.8% TD and 99.1% TD could be achieved. A number of SEM pictures have been produced. Then the mean ratio of pore diameters perpendicular and parallel to the direction of the applied electric field was estimated for a larger number of pores. The mean values obtained from four independent sampling procedures were 1.084 ± 0.068 and 1.152 ± 0.052 for the samples sintered at 1250 °C and 1260 °C, respectively. These results gave clear evidence that microwave sintering results in non-isotropic pore closure with mean pore aspect ratios significantly larger than one if samples were sintered in a linearly polarized microwave field.

Table I. Summary of process parameters for conventionally and microwave sintered TZ3YS samples and measured mean pore aspect ratios.

Material	Sintering temperature	Soak time	Sintering method	Sintered density	Mean pore aspect ratio	Av. number of eval. pores
	1230°C	3 days	MWS	5.91 g/cm^3	1.069 ± 0.063	125
	1230°C	3 days	MWS	5.74 g/cm^3	1.010 ± 0.002	176
	1230°C	3 days	MWS	5.53 g/cm^3	1.017 ± 0.092	128
TZ3YS	1250°C	7 days	MWS	5.99 g/cm^3	1.084 ± 0.068	137
	1260°C	4 days	MWS	6.02 g/cm^3	1.152 ± 0.052	49
	1270°C	1 day	MWS	5.81 g/cm^3	1.219 ± 0.056	256
	1300°C	1 day	MWS	6.03 g/cm^3	1.209 ± 0.045	48
	1400°C	1 h	CS	5.84 g/cm^3	1.081 ± 0.030	308

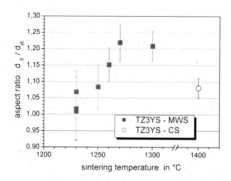

Fig. 6. Pore aspect ratio as a function of sintering temperature for TZ3YS.

Further sintering experiments were performed at lower as well as higher temperatures in the range from 1230 °C to 1300 °C. All process parameters and results on pore aspect ratio analysis are summarized in Table I. While at 1230 °C several days soak was necessary to achieve reasonable densities at 1300 °C one day soak was sufficient. Subsequent pore analysis revealed that with increasing sintering temperatures the mean aspect ratio of the pores was increasing (see Figure 6). This might be contradictory to the idea of using low sintering temperatures in order to reduce competing thermal effects. On the other hand for increasing sintering temperatures increased microwave power levels have to be applied, since energy losses from the sample by convection and radiation were increasing with temperature. But increased microwave power is equivalent to stronger electromagnetic fields in the sample volume that should result in stronger microwave effects if there are any. A comparative experiment was performed in a resistant heated lab kiln with a sintering temperature of 1400 °C in order to make sure that this effect was microwave driven and not a result of another effect. Microstructure analysis revealed a non-isotropic pore structure with an aspect ratio significantly larger than unity but significantly smaller than the value obtained by microwave sintering at lower sintering temperatures.

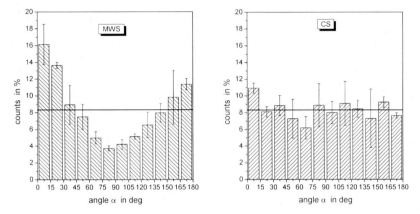

Fig. 7. Histogram showing orientation of pores in case of microwave (left) and conventionally (right) sintered TZ3YS.

For two samples, sintered to similar densities by microwaves at 1270 °C and conventionally at 1400 °C, the microstructure analysis by fitting ellipses was used as well, as described in the last chapter. Plotting this data into a histogram, showing the relative number of pores with the orientation of the large axis lying in different angle intervals with 15° in width resulted in graphs as shown in Figure 7. Assuming an even distribution of ellipses orientation the expected relative number of pores would be 8.33% if the half space from zero to 180° is divided into 12 equal intervals. The shown error bars were obtained by performing the evaluation procedure with the identical sample but by three different persons. This demonstrates the uncertainty of such an evaluation method. Nevertheless the results revealed exactly what has been expected according to the model proposed by Booske et al. [1]. In case of microwave sintering there was a distinct increase in the number of pores oriented perpendicular to the applied electric field, what corresponds to an angle near 0° or 180°, and a decrease in pores oriented parallel to it, what corresponds to an angle in the area of 90°. In contrast to this within the accuracy of this method in case of conventional sintering the angular distribution of the pores was even.

SUMMARY

During the last decade theoretical models of non-thermal microwave effects have been developed, based on the theory of ponderomotive driving forces in ionic crystals from K. Rybakov and V. Semenov [16]. A microstructural view of this model, in the vicinity of grain boundaries and residual pores in ceramic compacts, led to the prediction of a non-isotropic pore closure under the influence of linearly polarized electric fields [1]. However till now an experimental prove did not exist. This paper describes a possible method for experimental verification of such an effect. Ceramic compacts of nanosized 3-mol% yttria stabilized zirconia prepared by electrophoretic deposition technique were sintered in a single-mode TE_{103}

waveguide applicator in the maximum of the linearly polarized electric field. By variation of process temperatures the effective strength of the electric field within the sample could be varied. A subsequent investigation of the pore structure of the sintered compacts by scanning electron microscopy revealed strong evidence for the predicted non-thermal effects. If the pore aspect ratio was determined as the ratio of the maximum pore extension perpendicular to the applied electric field and parallel to it, average pore aspect ratios significantly larger than 1 were found with maximum mean values of about 1.2. This parameter was found to increase with increasing sintering temperatures, which is equivalent to increased applied microwave fields. Microstructure analysis by approximating isolated pores by ellipses revealed that in case of sintering in a linearly polarized microwave field the number of pores oriented perpendicular to the field direction is significantly increased. For the future further experiments are planned in order to improve the evidence of this non-thermal microwave effect.

ACKNOWLEDGEMENT

We kindly acknowledge Dr. Matthias Wolff, University of Saarbrücken, Germany for supply of the ceramic green bodies. The financial support from the Helmholtz-Gemeinschaft (VH-FZ-024) is gratefully acknowledged.

REFERENCES

[1] J.H. Booske R.F. Cooper, S.A. Freeman, K.I. Rybakov, and V.E. Semenov, "Microwave ponderomotive forces in solid-state ionic plasmas"; Physics of Plasmas, **5** [5] 1664-1670 (1998).

[2] W.H. Sutton, "Microwave firing of high alumina castables"; pp. 287-295 in Materials Research Society Symposium Proceedings Vol. 124, Microwave Processing of Materials. Edited by W.H. Sutton, M.H. Brooks, I.J. Chabinsky. Materials Research Society, Pittsburg, PA, 1988.

[3] M.A. Janney and H.D. Kimrey, "Diffusion Controlled Processes in Microwave-Fired Oxide Ceramics"; pp. 215-227 in Materials Research Society Symposium Proceedings Vol. 189, Microwave Processing of Materials II. Edited by W.B. Snyder, W.H. Sutton. Materials Research Society, Pittsburg, PA, 1991.

[4] J. Wilson and S.M. Kunz, "Microwave sintering of partially stabilized zirconia"; J. Am. Ceram. Soc. **71** [1] C40-C41 (1988).

[5] G. Link, L. Feher, M. Thumm, H.-J. Ritzhaupt-Kleissl, R. Böhme, and A. Weisenburger, "Sintering of advanced ceramics using a 30-GHz, 10-kW, cw industrial gyrotron"; IEEE Transactions on Plasma Science, **27** [2] 547-554 (1999).

[6] J. Cheng, J. Qiu, J. Zhou, and N. Ye, "Densification kinetics of alumina during microwave sintering"; pp. 323-328 in Materials Research Society Symposium Proceedings Vol. 269, Microwave Processing of Materials III. Edited by R.L. Beatty, W.H. Sutton, M.F. Iskander. Materials Research Society, Pittsburg, PA, 1992.

[7] M.A. Janney, H.D. Kimrey, M.A. Schmidt, and J.O. Kiggans, "Grain growth in microwave-annealed alumina"; J. Am. Ceram. Soc. **74** [7] 1675-81 (1991).

[8] R. Wroe and A.T. Rowley, "Evidence for a non-thermal microwave effect in the sintering of partially stabilised zirconia"; J. Mat. Science **31** [8] 2019-26 (1996).

9   G. Link, S. Takayama, and M. Thumm, "Critical assessment on temperature measurement and its consequences to observed sintering kinetics"; pp. 473-476 in Proc. of 9th Int. Conf. on Microwave and High Frequency Heating September 2003, Loughborough, UK. Edited by J. Binner Loughborough University 2003.

10 Link G., Rhee S., Thumm M., "Dilatometer system for investigations of sintering in a mm-wave oven." Proc. of the 36th Annual Microwave Symposium of the IMPI, San Francisco, April 18-21, pp. 23-26 (2001).

11 S.J. Rothman, "Critical assessment of microwave-enhanced diffusion"; pp. 9-18 in Materials Research Society Symposium Vol. 347 Microwave Materials Processing IV. Edited by M.F. Iskander, R.J. Lauf, W.H. Sutton. Materials Research Society, Pittsburgh, PA 1994.

12 Z. Fathi, D.E. Clark, and R. Hutcheon, "Surface modification of ceramics using microwave energy"; pp. 347-351 in Materials Research Society Symposium Proceedings Vol. 269, Microwave Processing of Materials III. Edited by R.L. Beatty, W.H. Sutton, M.F. Iskander. Materials Research Society, Pittsburg, PA, 1992.

13 M.A. Janney, H.D. Kimrey, W.A. Allen, and J.O. Kiggans, "Enhanced diffusion in sapphire during microwave heating"; J. Mat. Science 32 1347-55 (1997).

14 S.A. Freeman, J.H. Booske, and R.F. Cooper, „Novel method for measuring intense microwave radiation effects on ionic transport in ceramic materials" ; Rev. Sci. Instrum. 66 [6], 3606-3609 (1995).

15 M. Millert-Porada, "Microwave effects on spinodal decomposition"; pp. 403-409 in Materials Research Society Symposium Proceedings Vol. 430, Microwave Processing of Materials V. Edited by M.F. Iskander, J.O. Kiggans, J.-Ch. Bolomey. Materials Research Society, Pittsburg, PA, 1996.

16 K.I. Rybakov, V.E. Semenov, S.A. Freeman, J.H. Booske, and R.F. Cooper, "Dynamics of microwave-induced currents in ionic crystals"; Physical Review B, 55 [6] 3559-3567 (1997).

17 K.I. Rybakov and V.E. Semenov, "Possibility of microwave-controlled surface modification"; pp. 435-440 Materials Research Society Symposium Proceedings Vol. 430, Microwave Processing of Materials V. Edited by M.F. Iskander, J.O. Kiggans, J.-Ch. Bolomey. Materials Research Society, Pittsburg, PA, 1996.

18 K.I. Rybakov and V.E. Semenov, "Possibility of plastic deformation of an ionic crystal due to the nonthermal influence of a high-frequency electric field"; Phys. Rev. B 49 [1] 64-68 (1994).

19 A. Birnboim, J.P. Calame, and Y. Carmel, "Microfocusing and polarization effects in spherical neck ceramic microstructures during microwave processing"; J. App. Phys. 85 [1], 478-482 (1999).

20 A. Möbius and M. Mühleisen, "Vorrichtung zur räumlichen Justierung der Maxima und Minima von elektromagnetischen Feldern in einer Prozeßanlage zur thermischen Prozessierung von Materialien (apparatus for spacial adjustment of maxima and minima of electromagnetic fields wihtin a process equipment for thermal processing of materials)"; Patent DE-PS 10114022C1 (24.10.2002).

21 M. Sato et al., "Insulation blankets for microwave sintering of traditional ceramics"; pp. 277-285 in Ceramic Transactions Vol. 111, Microwaves: Theory and Application in Materials Processing V. Edited by D.E. Clark, J.G.P. Binner, D.A. Lewis, 2001.

22 O. Guillon, L. Weiler, E. Aulbach, and J. Rödel, "Anisotropic sintering of alumina thin films";
pp. 204-207 in Proceedings of the 4th International Conference on Science, Technology and

Applications of Sintering, Grenoble, France. Edited by D. Bouvard, August 29-September 1, 2005.

[23] R.K. Bordia, R. Zuo, O. Guillon, S.M. Salamone, and J. Rödel, "Anisotropic constitutive laws for sintering bodies"; Acta Materialia **54** 111-118 (2006).

[24] J. Tabellion and R. Clasen, "Electrophoretic deposition from aqueous suspensions for near-shape manufacturing of advanced ceramics and glasses – applications"; J. Mater. Sci., **39** 803-811 (2004).

THE EFFECT OF THE ELECTRICAL PROPERTIES ON THE PULSED ELECTRIC CURRENT SINTERING BEHAVIOR OF ZrO$_2$ BASED CERAMIC COMPOSITES

K. Vanmeensel, B. Neirinck, S.Huang, S. Salehi, O. Van der Biest, J. Vleugels

Department of Metallurgy and Materials Engineering (MTM), K.U.Leuven
Kasteelpark Arenberg 44
Heverlee, Belgium, 3001

ABSTRACT

The influence of the material's electrical properties on the pulsed electric current sintering (PECS) behavior of Y-ZrO$_2$ based ceramic composites is investigated in detail. The current and temperature distributions in the sintering compacts are investigated, using a developed finite element code, and their influence on the densification behavior is highlighted. Both zirconia ceramics and zirconia based ceramic composites with improved mechanical properties, as compared to traditionally hot pressed ceramics, could be obtained by PECS.

INTRODUCTION

During the last decade, the applicability of zirconia to induce toughening by the stress-induced transformation of the tetragonal to monoclinic ZrO$_2$ phase in the stress field of propagating cracks, a phenomenon known as transformation toughening[1,2], has been intensively investigated. Recent developments in zirconia composites are focused not only on the improvement of toughness, strength and hardness, but also on the possibility for mass production and manufacturing cost reduction. A successful approach is to incorporate electrically conductive reinforcements such as TiB$_2$[3], WC[4], ZrB$_2$[5], TiC[6], TiCN[7] and TiN[8] into the zirconia matrix. The incorporation of a certain content of these conductive reinforcements makes the composite electrically conductive enough to be machineable by electrical discharge machining (EDM), thus avoiding the expensive cutting and grinding operation for component shaping. Electrical resistivity threshold values for EDM are reported to be in-between 100 to 300 Ohm.cm[9,10]

Traditionally, these composites are densified by hot pressing[8]. This study investigates the possibility to process ZrO$_2$-TiCN nanocomposites by PECS. In our previous studies[11,12], it was pointed out that the changing thermal and electrical properties of a sintering ZrO$_2$-TiN (60/40) composite powder compact should be taken into account in order to be able to get a realistic idea of the temperature distribution in the tool and sample during PECS processing.

The goal of the present work is to process an electrically conductive ZrO$_2$-TiCN (60/40) (vol%) nanocomposite material by PECS within a few minutes, to investigate the densification behavior, as compared to traditional hot pressing, and to evaluate the mechanical and thermo-electrical properties. Using this information, the sinter set-up was optimised to improve the sintering homogeneity and reproducibility.

EXPERIMENTAL PROCEDURES

The commercial powders are yttria-free monoclinic ZrO$_2$ (Tosoh grade TZ-0, Tokyo, Japan), 3 mol % yttria stabilised zirconia (Daiichi grade HSY-3U, Japan) and TiCN (HTNMC grade, Hebei Sinochem, China). The Y$_2$O$_3$-stabiliser content was adjusted by mixing the appropriate amounts of monoclinic and yttria-stabilised zirconia powder [13].

The powder mixtures were ball milled in a multidirectional Turbula mixer (type T2A, Basel, Switzerland) in ethanol in a polyethylene container of 250 ml during 24 h at 60 rpm. 250 grams zirconia milling balls (Tosoh grade TZ-3Y, Tokyo, Japan) with a diameter of 15 mm were added to the container to break the agglomerates in the starting powder and to enhance powder mixing. The ethanol was removed after mixing using a rotating evaporator.

The dry powder mixture was sieved and inserted into a graphite die/punch set-up (inner diameter of 40 mm – outer diameter of 56 mm). All samples were densified using PECS (Type HP D 25/1, FCT Systeme, Rauenstein, Germany, equipped with a 250 kN uniaxial-press) in vacuum (~0.05 Pa) for 2 minutes at 1500°C under a load of 56 MPa, applying a heating rate of 200°C/min. Hot pressing (W100/150-2200-50 LAX, FCT, Rauenstein, Germany) was performed in vacuum at 56 MPa and 1500°C, applying a maximum heating rate of 50°C/min.

The density of the samples was measured in ethanol, according to the Archimedes method (BP210S balance, Sartorius AG, Germany). The Vickers hardness (HV_{10}) was measured on a hardness tester (Model FV-700, Future-Tech, Japan) with an indentation load of 98 N. The indentation toughness, K_{IC}, based on the crack length measurement of the radial crack pattern produced by Vickers HV_{10} indentations, was calculated according to the formula of Anstis et al.[14]. Vickers hardness profiles, containing at least 10 indentations, were obtained from polished cross-sections of 40 mm diameter discs. The elastic modulus (E) was measured using the resonance frequency method[15]. The resonance frequency was measured by the impulse excitation technique (Model Grindo-Sonic, J. W. Lemmens N.V., Leuven, Belgium). X-ray diffraction (Seifert 3003 T/T, Ahrensburg, Germany) analysis was used for phase identification and calculation of the transformability of the ZrO_2-based composites. The electrical resistance of the samples was measured according to the 4-point contact method using a Resistomat (TYP 2302 Burster, Gernsbach, Germany).

RESULTS AND DISCUSSION

In the first two paragraphs, the influence of the thermal and electrical properties of sintering Y-ZrO_2, TiCN, and ZrO_2-TiCN (60/40) composite powder compacts on the temperature and current distributions during PECS will be investigated. In the following paragraphs, the densification behavior of both monolithic Y-ZrO_2 and ZrO_2-TiCN (60/40) composite powder compacts during PECS will be investigated. Both their mechanical, microstructural as well as electrical properties will be investigated in detail.

Thermal and electrical properties of sintering ceramic powder compacts

In order to be able to calculate the electrical properties of sintering powder compacts during PECS, two sintering experiments were performed to determine the densification behavior of both 2Y-ZrO_2 and ZrO_2-TiCN (60/40) composite powder compacts. The temperature and pressure cycles, shown in Figure 1, were used throughout both experiments. The open symbols indicate at which temperature/pressure combination the sintering cycle was interrupted in order to obtain more detailed information on the densification behavior of the composites.

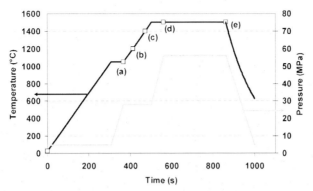

Fig. 1. *PECS sintering cycle, used during the finite element simulations. A heating rate of 200°C/min was applied, whereas a pressure of 56 MPa was applied during the final dwell period at 1500°C.*

A typical PECS sintering cycle consisted of six segments: 1 – application of a constant voltage in combination with a pressure of 7 MPa until a temperature of 450°C was recorded by a central pyrometer focusing on the bottom of a borehole inside the upper punch of the PECS tool set-up. 2 – heating segment from 450 to 1050°C applying a constant pressure of 7 MPa and a constant heating rate of 200°C/min. 3 – dwell period at 1050°C (1') in order to increase the pressure from 7 to 28 MPa. 4 – heating segment from 1050 to 1500°C applying a pressure of 28 MPa and a constant heating rate of 200°C/min. 5 – dwell period at 1500°C (6') while the pressure was increased from 28 to 56 MPa during the first minute of the dwell segment. 6 – free cooling by switching off the current. The evolution of the theoretical density of both the Y-ZrO$_2$ and ZrO$_2$-TiCN (60/40) composite powder compacts is summarized in Table 1. The theoretical density of 2Y-ZrO$_2$ and TiC$_{0.5}$N$_{0.5}$ were taken as 6.03 and 5.18 g/cm^3, respectively.

Table 1. *Evolution of the relative density (% RD) of sintering 2Y-ZrO$_2$ and 2Y-ZrO$_2$-TiC$_{0.5}$N$_{0.5}$ (60/40) composite powder compacts as function of the temperature / pressure combinations used during the interrupted sintering experiments, as shown in Figure 1.*

Temperature (°C) / Pressure (MPa)	2Y-ZrO$_2$ (% RD)	2Y-ZrO$_2$ – TiCN (60/40) (% RD)
1050 / 28	38.2	53.7
1200 / 28	52.9	73.0
1400 / 28	91.4	94.1
1500 / 56 (1')	97.0	95.8
1500 / 56 (6')	98.1	97.3

When the densification behavior of the Y-ZrO$_2$ and the Y-ZrO$_2$-TiCN (60/40) composite powder compacts is compared (Table 1), it can be observed that the green packing of the composite is much better compared to the Y-ZrO$_2$ powder compact. In both cases, most densification takes place during the second heating stage between 1050 and 1500°C. In case of the composite powder compact, closed porosity is obtained during this stage, whereas the Y-

ZrO_2 compact reaches closed porosity during the dwell period at 1500°C. Contrary to ZrO_2-TiN (60/40) composite powder compacts, containing micrometer-sized TiN particles[12], the PECS densification behavior of the composite material is not retarded by the presence of the secondary nanometer-sized $TiC_{0.5}N_{0.5}$ phase. This can be attributed to: a) the higher green density of the ZrO_2-TiCN (60/40) composite powder compacts, as compared to ZrO_2-TiN (60/40) powder compacts b) the fine grain size of the secondary $TiC_{0.5}N_{0.5}$ powder, hereby enhancing the sinterability and c) the sintering activation of the electrical field when electrically conductive powder compacts are densified by PECS.

This last statement will be investigated in the following paragraph by comparing the densification behavior of a ZrO_2-TiCN (60/40) composite powder compact during pulsed electric current sintering and traditional hot pressing (HP). In order to do this, the electrical properties of both fully dense and sintering $2Y$-ZrO_2-$TiC_{0.5}N_{0.5}$ (60/40) composite powder compacts were calculated as function of the sintering temperature, using the densifcation behavior, shown in Table 1. The electrical properties of the different components were taken from literature[16,17], while the composite properties and the influence of porosity were included using theoretical mixture rules[18,19]. The results are summarized in Figure 2. Since a percolation type of mixture rule was used, assuming a homogeneous dispersion of spherical secondary phase particles, the electrical properties of the sintering composite powder compact change very quickly in the 1200-1300°C temperature range. Before percolation, the sintering composite powder compact will act as an insulator, forcing the current to flow from the upper punch into the die, while after reaching percolation, the current can also pass through the sintering powder compact.

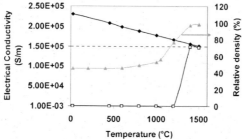

Fig. 2. *Evolution of the relative density (Δ) and the electrical conductivity of a fully dense ($\blacklozenge$) as well as a sintering ($\square$) ZrO_2-$TiC_{0.5}N_{0.5}$ (60/40) composite powder compact as function of the sintering temperature. Calculations based on the densification behavior, summarized in Table 1.*

Temperature and current distributions during PECS

The calculated electrical properties of the sintering ZrO_2-TiCN (60/40) composite powder compacts were inserted into a previously developed finite element code[20] in order to calculate the temperature and current distributions during PECS. Before percolation, the current is forced from the upper punch, through the die, to the lower punch (Figure 3-a). The powder compact is heated by conduction of Joule heat from the die and punches (Figure 3-b). After reaching percolation, direct Joule heating of the powder compact takes place (Figure 3-c) and the die is cooled (Figure 3-d) due to radiation heat losses. A radial temperature gradient is present in the sample (Figure 6-a).

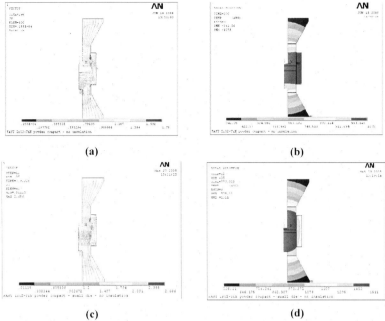

Fig. 3. *Current density (a-c) and temperature distribution (b-d) before (a-b) and after (c-d) percolation takes place during PECS of a ZrO₂-TiCN (60/40) composite powder compact.*

In order to investigate whether the current influences the densification behavior of both 2Y-ZrO₂ and ZrO₂-TiCN (60/40) composite powder compacts, interrupted PECS and hot pressing (HP) experiments, using identical temperature (50°C/min heating rate – 5' dwell time at 1500°C) and pressure cycles (56 MPa – 2 step pressure application with 1' dwell at 1050°C), were performed. The evolution of the relative density of the partially sintered powder compacts is summarized in Table 2 as function of the sintering temperature.

Table 2. *Evolution of the relative density of sintering 2Y-ZrO₂ and 2Y-ZrO₂-TiC$_{0.5}$N$_{0.5}$ (60/40) composite powder compacts as function of the temperature / pressure combinations used during the interrupted PECS / HP experiments, applying a heating rate of 50°C/min.*

Temperature (°C)	2Y-ZrO₂ (HP) (%)	2Y-ZrO₂ (PECS) (%)	ZrO₂-TiCN (HP) (%)	ZrO₂-TiCN (PECS) (%)
1050	44,0	41,8	55,1	56,1
1200	68,5	66,2	60,6	76,6
1400	97,8	98,4	81,9	96,9
1500 (1')	99,1	99,1	97,6	97,3
1500 (6')	99,2	99,3	98,6	98,2

The densification behviour of a 2Y-ZrO₂ compact is similar, independent of whether the powder compact was densified by HP or PECS. In case that the densification behavior of a ZrO₂-TiCN (60/40) compact using PECS is compared with that inside a hot press, PECS enhances the densification behavior in the intermediate stage of sintering, when a percolating network is formed inside the sintering powder compact. Due to the fact that the secondary TiCN phase was not perfectly dispersed into the Y-ZrO₂ matrix, as shown in Figure 7-a, percolation might even occur at lower temperatures, explaining the enhanced densifcation behavior in between 1050 and 1200°C. Therefore it can be concluded that the presence of the electrical field enhances densification in the intermediate sintering stage. Furthermore, closed porosity is reached already at 1400°C, allowing the sintering temperature to be reduced, as will be discussed in the following paragraphs. When the results in Tables 1 and 2 are compared, it is observed that a higher heating rate retards densification, both in case that of Y-ZrO₂ and the ZrO₂-TiCN (60/40) composites.

PECS of yttria stabilized zirconia (Y-TZP) ceramics

The microstructural and mechanical properties of 2Y-ZrO₂ materials, densified by HP and PECS, were compared. All Y-ZrO₂ samples, densified at 1400 and 1500°C, were thermally etched in oxygen at 1350°C for 20 minutes, while secondary electron (SE) micrographs of their microstructures are shown in Figure 4. The transformability of the Y-ZrO₂ matrix could be calculated from the X-ray diffraction (XRD) patterns of smoothly polished and fracture surfaces, using the method proposed by Toraya[21]. The mechanical properties are summarized in Table 3.

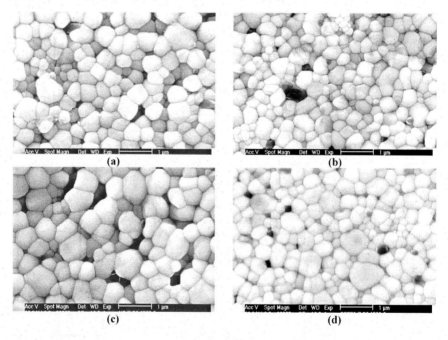

(a)   (b)

(c)   (d)

e)   f)

Fig. 4. *Representative secondary electron (SE) micrographs of thermally etched 2Y-ZrO₂ ceramics, densified by either hot pressing (HP) (a-c-e) or pulsed electric current sintering (PECS). All ceramics were subjected to identical sintering cycles, applying a heating rate of 50°C/min up to 1500°C (6') and a pressure of 56 MPa. The sintering cycles were interrupted at 1400°C (a-b) and at 1500°C after 1 minute (c-d) and 6 minutes (e-f) dwell time. light grey: Y-ZrO₂ – dark: Al₂O₃.*

Table 3. *Comparison of mechanical properties of hot pressed(HP) and pulsed electric current sintered (PECS) 2Y-ZrO₂ ceramics as function of sintering temperature*

Temperature (°C)	Young's modulus (GPa)	Vickers hardness (GPa)	Fracture toughness (MPa.m$^{1/2}$)	Transformability ZrO₂ (vol %)	3-pt bending strength (MPa)
HP 1400	204	11.9 ± 0.3	6.5 ± 0.3	43	-
HP 1500-1	208	12.4 ± 0.1	7.2 ± 0.1	50	1167 ± 67
HP 1500-6	210	12.2 ± 0.1	6.8 ± 0.2	47	-
PECS 1400	206	11.9 ± 0.2	6.8 ± 0.1	50	-
PECS 1500-1	209	12.4 ± 0.1	7.4 ± 0.6	63	1271 ± 73
PECS 1500-6	208	12.4 ± 0.1	7.7 ± 0.3	62	-

When the SEM micrographs of the HP and PECS sintered Y-ZrO₂ ceramics are compared, it is clear that the average Y-ZrO₂ grain size is always larger in case that the material was hot pressed. Figure 5 compares the cumulative grain size distribution of the materials that were densified at 1500°C for 1 minute. Using Image Pro Plus software, the average grain size of the hot pressed material was determined to be 0.486 ± 0.194 μm. while the PECS sintered material had an average grain size of 0.356 ± 0.170 μm. In order to determine whether this was related to an intrinsic characteristic of the PECS process or to the slower cooling rate inside the hot press (20°C/min vs 200°C/min in case of PECS), a PECS experiment with a controlled cooling rate of 20°C/min was performed. The average grain size of this material was 0.456 ± 0.181 μm. so that the difference in cooling rate is the main reason for the finer microstructure in case that the 2Y-TZP material is densified by PECS. In case that the PECS material was cooled at a 20°C/min, Ostwald ripening of the larger grains took place, hereby removing the smaller grains that were still observed in case that a high cooling rate was applied (Figure 4–f), shifting the cumulative grain size distribution to the right.

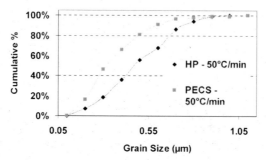

Fig. 5. *Comparison of the cumulative grain size distributions of the 2Y-ZrO₂ materials hot pressed (♦) and pulsed electric current sintered (■) at 1500°C for 1 minute.*

All 2Y-ZrO₂ materials had comparable densities, Young's moduli and Vickers hardness values. The fracture toughness and transformability of the PECS sintered specimens was slightly higher, due to their bimodal grain size distribution. Due to the combination of the smaller grain size and the higher transformability of the PECS sintered specimens, it is believed that the fracture strength can be increased, which was experimentally verified in a 3-point bending test, comparing the strength of the materials that were hot pressed and pulsed electric sintered at 1500°C for 1 minute (Table 3).

PECS of Y-TZP based ceramic composites

The finite element simulations, shown in Figure 3, indicate that a radial temperature gradient exists inside the sintering ZrO_2-TiCN (60/40) composite powder compacts when the sintering temperature of 1500°C is reached. Figure 6-a shows that the sample edge is 135°C cooler, as compared to the sample centre. In order to improve the sintering homogeneity, the graphite die was surrounded with thermal carbon felt insulation so that the radiation heat losses could be reduced. Figure 6-b indicates the effectiveness of this measure since the radial temperature gradient in the sample is reduced to 30°C.

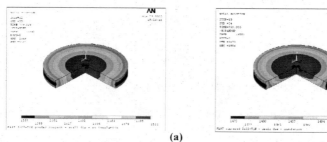

(a)   (b)

Fig. 6. *Radial temperature distribution inside a sintering ZrO_2-TiCN (60/40) composite material during the dwell period at 1500°C. (a) free radiating die (b) die surrounded with thermal carbon felt insulation.*

Furthermore, it was observed that the presence of an electrical current during PECS enhanced the intermediate sintering stage, as shown in Table 3. Representative backscattered electron micrographs (BSE) of the ZrO_2-TiCN (60/40) composite material, sintered at 1500°C for 1 minute, are shown in Figure 7. It can be observed that, although the TiCN particles are not homogeneously distributed inside the Y-ZrO_2 matrix, coarsening of the secondary carbonitride phase is limited. The material combines a Young's modulus of 231 GPa with a Vickers hardness of 1352 ± 29 kg/mm^2 and a fracture toughness of 6.0 ± 0.5 MPa.m$^{1/2}$. The transformability of the Y-ZrO_2 matrix phase was 23 %, while a bending strength of 1156 ± 41 MPa was obtained. As compared to the Y-ZrO_2 ceramics, the composite materials are more brittle and exhibit a lower bending strength, but a higher Vickers hardness. The main reason for the decreased fracture toughness and bending strength is the presence of a substoichiometric Ti_2O_3 oxide layer on the outer surface of the TiCN particles, as can be observed from the XRD spectrum, shown in Figure 8. Titanium atoms can dissolve into the tetragonal Y-ZrO_2 lattice, hereby stabilising the tetragonal phase and inhibiting its transformability[22].

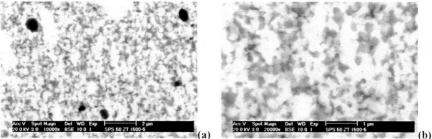

(a)   (b)

Fig. 7. *Back scattered electron micrographs (BSE) of the ZrO_2-TiCN (60/40) composite material pulsed electric current sintered at 1500°C for 1 minute. white: Y-ZrO_2 – grey: TiCN – black: Al_2O_3.*

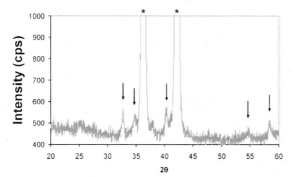

Fig. 8. *X-ray diffraction pattern of the nanometer-sized $TiC_{0.5}N_{0.5}$ starting powder, revealing the presence of a substoichiometric Ti_2O_3 impurity (↓). *: $TiC_{0.5}N_{0.5}$*

CONCLUSIONS

The pulsed electric current sintering behavior of Y-TZP based ceramic materials was investigated. It was observed that the electrical properties of the sintering powder compacts influenced the temperature and distributions during PECS as well as the densification behavior of the materials themselves.

The densification behavior of a 2Y-TZP material in PECS is similar to that of hot pressing (HP). However, a finer final grain size can be obtained by PECS, due to the much higher cooling rate, resulting in an increased transformability, fracture toughness and bending strength.

When Y-ZrO$_2$-TiCN (60/40) composite materials is densified by PECS, an adjusted sintering set-up has to be used to minimize the radiation heat losses from the die and the radial temperature gradient in the sintering composite sample. The presence of an electrical current enhanced the densification behavior during the intermediate sintering stage, as compared to traditional hot pressing. An electrically conductive composite material with a nanometer-sized microstructure and acceptable mechanical properties could be obtained within minutes.

The PECS technique allows full densification with minimal grain growth, a critical issue to obtain nanosized (<100 nm average grain size) materials with superior mechanical properties.

ACKNOWLEDGMENTS

This work was supported by the GROWTH program of the Commission of the European Communities under project contract no. G5RD-CT2002-00732 as well as by the Framework 6 Program under project No. STRP 505541-1.

REFERENCES

[1] R.C. Garvie, R. H. J Hannink, and R.T. Pascoe, "Ceramic steel?", *Nature*, **258**, 703-704 (1975).

[2] M. Rühle, and A.G. Evans, "High toughness ceramics and ceramic composites", *Prog. Mater. Sci.*, **33**, 85-167 (1989).

[3] B. Basu, J. Vleugels, and O. Van Der Biest, "Toughness optimisation of ZrO$_2$-TiB$_2$ composites", *Key Engineering Materials*, **206-213**, 1177-1180 (2002).

[4] G. Anné, S. Put, K. Vanmeensel, D. Jiang, J. Vleugels, and O. Van der Biest, "Hard, tough and strong ZrO$_2$-WC composites from nanosized powders", *J. Eur. Ceram. Soc.*, **25**, 55-63 (2005).

[5] B. Basu, J. Vleugels, and O. Van der Biest, "Development of ZrO$_2$-ZrB$_2$ composites", *Journal of Alloys and Compounds*, **334**, 200-204 (2002).

[6] J. Vleugels, and O. Van der Biest, "Development and characterization of Y$_2$O$_3$-stabilized ZrO$_2$ (Y-TZP) composites with TiB$_2$, TiN, TiC, and TiC$_{0.5}$N$_{0.5}$", *J. Am. Ceram. Soc.*, **82(10)**, 2717 -2720 (1999).

[7] K. Vanmeensel, K.Y. Sastry, A. Laptev, J. Vleugels, and O. Van der Biest, "Microstructure and mechanical properties of spark plasma sintered ZrO$_2$-Al$_2$O$_3$-TiC$_{0.5}$N$_{0.5}$ nanocomposites", *Solid State Phenomena*, **106**, 153-160, (2005).

[8] S. Salehi, O. Van der Biest and J. Vleugels, "Electrically conductive ZrO_2–TiN composites", *J Eur Ceram Soc*, **26 (15)**, 3173–3179 (2006).

[9] König, W., Dauw, D.F., Levy, G., U. Panten, EDM - future steps towards the machining of ceramics, Annals of the CIRP, 1988, 37, 623-631.

[10] R.F. Firestone, Ceramic Applications in Manufacturing, SME, Michigan, 1988, 133.

[11] K. Vanmeensel, A. Laptev, O. Van der Biest and J. Vleugels, Field assisted sintering of electro-conductive ZrO_2-based ceramics, *J Eur Ceram Soc* 27 (2007) (2–3), pp. 979–985

[12] K. Vanmeensel, A. Laptev, O. Van der Biest and J. Vleugels."The influence of percolation during pulsed electric current sintering of ZrO_2-TiN powder compacts with varying TiN content." *Acta Materialia*, In Press, Corrected Proof.

[13] B. Basu, J. Vleugels and O. Van der Biest. "Toughness tailoring of yttria-doped zirconia ceramics." *Materials Science and Engineering*, A 380(1-2), 215-221 (2004).

[14] G.R. Anstis, P. Chantikul, B.R. Lawn and D.B. Marshall, "A critical evaluation of indentation techniques for measuring fracture toughness", *J. Am. Ceram. Soc.* **64**, 533–538 (1981).

[15] G. Roebben, B. Basu, J. Vleugels, J. Van Humbeeck and O. Van der Biest, "The innovative impulse excitation technique for high-temperature mechanical spectroscopy", *J. Alloys Compd.*, **310**, 284–287 (2000).

[16] W. Lengauer, S. Binder, K. Aigner, P. Ettmayer, A. Guillou and J. Debuigne *et al.*, Solid state properties of group IVb carbonitrides, *J Alloys Compounds* 217 (1995), pp. 137–147

[17] A.W. Weimer, Carbide, nitride and boride materials synthesis and processing, Chapman & Hall, New York (1997) Appendix D, pp. 653–654.

[18] G. Bánhegyi, Comparison of electrical mixture rules for composite, *Colloid Polymer Sci*, **264**, 1030–1050 (1986).

[19] R. Landauer, The electrical resistance of binary metallic mixtures, *J Appl Phys*, **33**, 779–784 (1952).

[20] K. Vanmeensel, A. Laptev, J. Hennicke, J. Vleugels and O. Van der Biest, "Modeling of the temperature distribution during field assisted sintering", *Acta Mat*, **53**, 4379–4388, (2005).

[21] H. Toraya, M. Yoshimura and S. Somiya, "Calibration Curve for Quantitative Analysis of the monoclinic-tetragonal ZrO_2 System by X-ray Diffraction", *J. Am. Ceram. Soc.*, **67**, C119-C121 (1984).

[22] S.N.B. Hodgson and J. Cawley, "The effect of titanium oxide additions on the properties and behaviour of Y-TZP." *Journal of Materials Processing Technology* **119**(1-3): 112-116 (2001).

SMART PROCESSING DEVELOPMENT ON 3D CERAMIC STRUCTURES FOR TERAHERTZ WAVE APPLICATIONS

Y.Miyamoto, W.Chen, H.Kanaoka, and S.Kirihara

Smart Processing Research Center, Joining and Welding Research Center, Osaka University, 11-1 Mihogaoka, Ibaraki, Osaka 567-0047, Japan

ABSTRACT

Smart processing to fabricate 3D ceramic structures such as micro photonic crystals and fractals has been developed using micro-stereolithography and sintering techniques. Micro photonic crystals with a diamond structure of ceramic/resin composites were formed from ceramic paste of TiO_2, SiO_2, or Al_2O_3 particles dispersed in photo curable resin by micro-stereolithography. Dense ceramic photonic crystals were obtained by dewaxing and sintering the performs of ceramic/resin photonic crystals. These micro photonic crystals showed deep and complete photonic band gaps in a terahertz range at around 300GHz to 500GHz. Terahertz waves could be localized in a micro-cavity designed in a ceramic photonic crystal. When a 3D micro fractal structure of Menger sponge cube is incorporated in the photonic crystal, we can expect the enhancement of localization in this dual structure.

INTRODUCTION

Smart processing is a new concept of materials processing to achieve excellent performance or new functions by supplying necessary energy and resources only to required parts. [1] We are developing such a smart processing for ceramics by employing stereolithography of a CAD/CAM system. It is possible to freely design and fabricate 3D complex structures of ceramics such as photonic crystals and fractals without mold by combining stereolithography and sintering techniques.

The photonic crystal is a dielectric material with the periodic structure which can totally reflect electromagnetic waves with the wavelength similar to the periodicity due to Bragg diffraction and form photonic band gap similar to electronic band gap in semiconductors. [2,3] Research and development on photonic crystals are increasing aiming to apply to advanced photonic or electromagnetic waveguides, circuits, filters, cavities, laser, antennas, absorbers, and others. [4,5]

While, the photonic fractal is a new dielectric material with the self-similar structure which can localize and confine electromagnetic waves with specific wavelengths associated with the fractal geometry and dielectric constant of component material. [6-8] We are investigating on the

79

localization mechanism of electromagnetic waves and the capability of applications to electromagnetic absorber, filter, cavity, antenna and others.

Recently, a micro-sterolithographic machine has been developed in collaboration with companies. [9,10] This micro-stereolithography enables to fabricate micrometer scale structures of ceramics in 3D complex forms. We are fabricating ceramic photonic crystals and fractals by using micro-stereolithography and successive sintering aiming to apply to terahertz wave devices.[11,12] Terahertz wave is the intermediate electromagnetic wave between radio wave and light. Its wavelength and frequency ranges in 3mm to 30 μm, and 100GHz to 10 THz, respectively. Recent developments in generation and detection of terahertz waves is leading to new technologies for imaging, communications, medical and chemical detections including gun powders and drags. [13] However, it is necessary to develop micro devices to control terahertz waves such as filters, cavities, waveguides, antenas for their wide applications.

We report on this new ceramic processing using micro-stereolithography and sintering as a smart processing and fabrication of 3D micro photonic crystals and fractals of ceramics as well as their terahertz wave properties.

MICRO-STEREOLITHOGRAPHY

Sterelithography is known as a rapid prototyping to form complex 3D objects layer by layer using laser scanning to liquid photo-curable resin with a CAD/CAM system. [14] It is used for fabrication of prototyping objects in motor, machine, electronic industries and medical uses. The conventional stereolithographic machine scans a laser beam from a fixed point by changing the reflection angle of a galvano mirror device so that the projected area is extended and the resolution of exposure is limited. While, micro-stereolithography (Acculus SI-C1000, D-MEC, Tokyo, Japan) employs a DMD (Digital Micro-mirror Device) as a dynamic mask as shown in Fig.1.

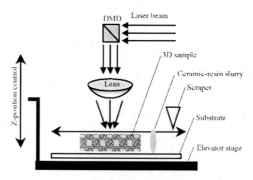

Fig.1   Schematic illustration of micro-stereolithography

DMD is an optical device assembled by micro-mirrors of 14 μm in edge length. Each mirror is controlled following the sliced CAD data. A 3D figure designed using the CAD program (Toyota Caelum Co. Ltd., thinkdesign ver.7.0, Aichi, Japan) is converted into STL file of a rapid prototyping format, sliced and compiled to a set of thin sections by slice software (Magics 9.9, Materialise Co.Ltd., Leuven, Belgium). A two dimensional figure is projected with a 405 nm laser using a DMD and an objective lens. During micro-stereolithography fabrication, the mixed paste of nanometer sized ceramic particles with a liquid photo curable resin was squeezed out from a dispenser by controlled air pressure and coated as thin as 2~20μm in thickness on a previously formed layer by moving a knife edge. Figure 2 shows the resolution of micro-stereolithography for a paste of 40vol.% TiO_2-acrylic resin with a laser intensity of 1000 mJ/cm^2 and layer thickness of 5 μm. Compared to the designed samples, the polymerized parts expanded about 10 μm. The cured width depends on the spot size of a focused laser beam and the spreading of laser radiation by side-scattering from ceramic particles. The curing depth is limited to 10 μm due to scattering by TiO_2 with high refractive index of ~2.5. This problem can be solved by reducing the layer thickness to 5 μm because 2~3 μm in thickness per layer can be achieved by micro-stereolithography.

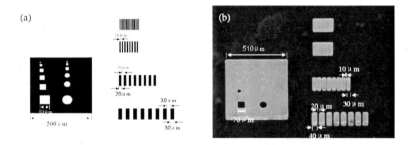

Fig.2 (a) CAD patterns for feature resolution check. Dark regions will be irradiated by laser. (b) Formed patterns with 40 vol.% TiO_2/resin by micro-stereolithogaphy.

MICRO PHOTONIC CRYSTALS AND TERAHERTZ WAVE PROPERTIES

Diamond structure is known as an ideal structure of photonic crystal because it can form a perfect band gap opened for all directions as illustrated in Fig.3. [15] It is difficult to fabricate such a complex structure by conventional methods. We have fabricated various diamond structures of ceramics/resin composites by using micro-stereolithography. Figure 4 (a) and (b) show a micro photonic crystal consisting of 8x8x4 diamond cells composed of 40 vol.% TiO_2-acrylic resin. The lattice constant is 500 μm. No large pore and crack were found. It is possible to convert this ceramic/resin object to ceramic one by dewaxing at 600 °C and sintering at 1400 °C in air. Figure 4 (c) is a sintered TiO₂ photonic crystal with a lattice constant of 380 μm. The relative density and the linear shrinkage ratio is 92% and 25%, respectively.

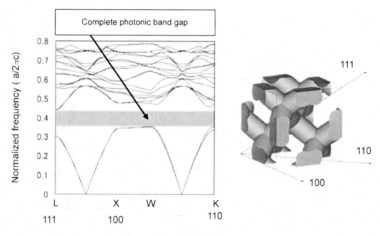

Fig.3 Schematic illustrations of a unit cell and photonic band diagram for a dielectric diamond structure.

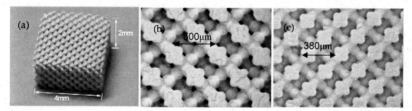

Fig.4 (a) Digital OM images of photonic crystals with diamond structures composed of 40 vol.% TiO₂/resin composite. (b) Top view at (100) plane. (c) Top view of a sintered photonic crystal.

Figure 5 (a) and (b) shows transmission spectra for photonic crystals composed of the TiO₂/resin composite and the TiO₂ ceramic, respectively, which were measured by using terahertz time domain spectrometer (J-Spec2001/ou, Advanced Infrared Spectroscopy Co.Ltd., Tokyo, Japan). A sharp band gap was observed between 280GHz~360GHz for a photonic crystal of the TiO₂/resin composite along the Γ-L <100> direction. The measured dielectric constant of the

TiO₂/resin composite was about 16 in this frequency range. The crystalline phase of TiO₂ was mainly anatase. In case of a TiO₂ photonic crystal, the attenuation of transmission was very large due to high absorption. The measured dielectric constant of the TiO₂ ceramic sintered was increased to about 100 because the crystalline phase was converted from anatase to rutile, but the absorption due to dielectric loss was increased. Figure 6 (a) and (b) compares two band diagrams for a TiO₂/resin photonic crystal and a TiO₂ one. They were calculated using measured dielectric constants and lattice parameters including the aspect ratios of lattice rods by a plane wave expansion method.[16] The measured photonic band gaps exist in the corresponding calculated band gaps.

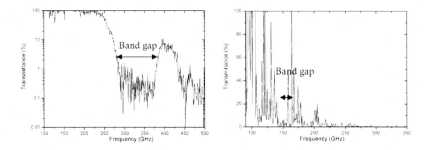

Fig.5 Transmission spectra of terahertz waves along theΓ–X <100> direction for micro photonic crystals with diamond structures of (a) 40 vol.% TiO₂/resin, (b) sintered TiO₂.

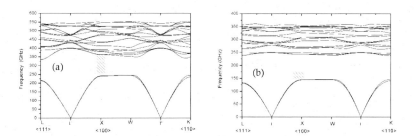

Fig.6 Calculated photonic band diagrams for diamond structures of (a) 40 vol.% TiO₂/resin, (b) sintered TiO₂. The measured photonic band gaps were shadowed.

The dielectric loss of many ceramic materials increases in terahertz frequency range. This is problem for application of ceramics. Table I summarizes transmission amplitudes of some sintered ceramics which were measured at 400 GHz. The SiO_2 shows a good transmission ability. Therefore, we fabricated a SiO_2 photonic crystal with a diamond structure from a precursor of a 50vol.% SiO_2 dispersed in acrylic resin by dewaxing at 600 °C and successive sintering at 1400 °C. The sintered SiO_2 photonic crystal is shown in Fig. 7. The relative density was 99% and the shrinkage ratio 26%. The dielectric constant of SiO_2 was about 3.5. The transmission spectrum of this sample along the Γ-L <111> direction is shown in Fig.8. It shows high transmission rate up to 700 GHz and a deep band gap between 400GHz and 520GHz. A complete photonic band gap was observed between 490GHz and 510 GHz.

Table 1 Transmission rate of terahertz wave at 400 GHz for various ceramics sintered.

Sintered Ceramics	SiO_2	ZTA	Al_2O_3	ZrO_2	TiO_2	$BaTiO_3$
Transmission rate %	90	65	50	15	3	0.01

(The thickness of samples is 1.5 mm.)

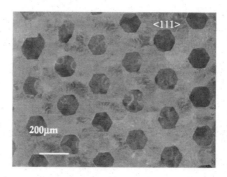

Fig.7 Digital OM image of a micro photonic crystal composed of sintered SiO_2.

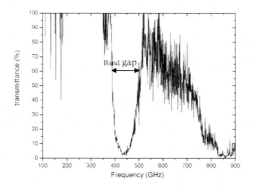

Fig.8 Transmission spectrum of terahertz waves along theΓ-L <111> direction for a micro photonic crystal with a diamond structure of sintered SiO$_2$.

LOCALIZATION OF TERAHERTZ WAVES IN MICRO PHOTONIC CRYSTAL AND FRACTAL

It is possible to localize terahertz waves in a defect of periodic structure such as a cavity designed in a photonic crystal. When a waveguide is designed, terahertz waves in a complete band gap can propagate along the waveguide by repeating total reflections with the surrounding diamond structure.

Fig.9   SEM image of diamond lattice of Al$_2$O$_3$ incorporating a micro-cavity defect.

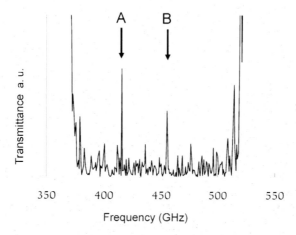

Fig.10 Transmission spectrum of terahertz waves along theΓ–X <100> direction for an Al₂O₃ photonic crystal with a micro cavity defect.

Figure 9 shows an Al₂O₃ photonic crystal in which a micro cavity with the unit cell size is designed. Figure 10 shows a transmission spectrum of this sample alongΓ-X <100> direction where two localized modes, A and B appeared at 420 GHz and 460 GHz in the band gap, respectively. The electric field intensity profiles at these localized frequencies in the structure were simulated by using TLM (Transmission Line Modeling) code and demonstrated in Fig. 11. The mode A localizes with a half wave oscillation having a node at the center in the cavity, while the mode B is oscillating along the diagonal direction with a half wave having an antinode in the center of cavity which is weaker than the mode A.

Photonic fractal has the self-similar structure of dielectric media which can localize or confine electromagnetic waves due to resonances in the structure. [17] We have fabricated 3D fractal structures called Menger Sponge and discovered the strong localization of electromagnetic waves. [4] Menger sponge is a cubic fractal which is formed by dividing each edge of the initial cube into three smaller equivalent segments, so that the cube is divided into 27 smaller identical cubes. After that, 7 smaller cubes at the face and body centers are taken away and 20 smaller cubes remain. This structure is called as the stage 1 Menger sponge. When this division and extraction process is repeated to the remaining 20 smaller cubes, we can obtain the stage 2 Menger sponge with large and small square channels crossing each other.

Figure 12 shows a stage 2 Menger sponge embedded in a photonic crystal of Al₂O₃. This multi-structure aims dual effects of the localization due to the resonance in the self-similar structure with a frequency designed in the band gap of photonic crystal. Figure 13 shows the transmission spectrum where a sharp localized mode appears at 416GHz in the band gap. This localization frequency agrees well with the localization frequency calculated for the stage 2 Menger-sponge using an empirical equation [17] and the simulated frequency using Transmission Line Modeling method. We are investigating the enhancement of localization by designing various multi-structures of periodic and self-similar ones.

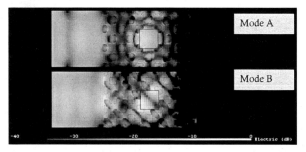

Fig.11 Electric field intensity profiles of localized waves in a micro cavity defect. (a) mode A at 420 GHz. (b) mode B at 460 GHz.

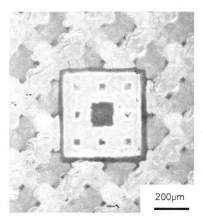

Fig.12 A dual structure of a micro Menger sponge fractal (stage 2) incorporated in a diamond structure of Al₂O₃.

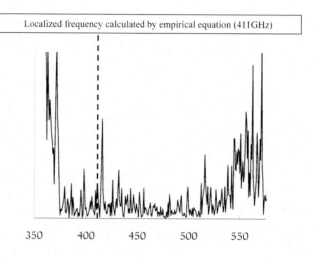

Fig.13 Transmission spectrum of terahertz waves along the $\Gamma-X$ <100> direction for a dual structure of a micro Menger sponge fractal (stage 2) incorporated in a diamond structure of Al_2O_3.

SUMMARY

Smart processing for ceramics using micro-stereolithography and sintering has been demonstrated in fabrication of 3D micro phonic crystals and fractals. Micro photonic crystals with a diamond structure of ceramic/resin composite could be formed directly from the paste of TiO_2, SiO_2, or Al_2O_3 particles dispersed in a photo curable resin by micro-stereolithography. By dewaxing and sintering of the ceramic/resin photonic crystals at optimized heating conditions, dense ceramic photonic crystals could be obtained. These micro photonic crystals composed of ceramics and ceramics/resin showed deep and complete photonic band gaps in terahertz wave range at around 300GHz to 500GHz. It is possible to localize terahertz waves in defect such as a micro cavity designed in a photonic crystal. When the fractal structure such as Menger sponge is incorporated in a photonic crystal, we can expect to enhance the localization in this dual structure.

ACKNOWLEDGEMENT

This work is supported by the Grant for the 21st Century's COE Program "Center of Excellence for Advanced Structural and Functional Materials Design", and by the Grant-in-Aid for Scientific Research (S) No. 171067010, both under the auspice of the Ministry of Education, Culture, Sports, Science and Technology, Japan.

REFERENCES

[1] Y.Miyamoto, S.Kirihara, S.Kanehira, M.W.Takeda, K.Honda, and K.Sakoda, "Smart Processing Development of Photonic Crystals and Fractals," *Int. J. Appl. Ceram. Technol.* 1, 40-48 (2004).

[2] E.Yablonovitch, "Photonic Band-Gap Structures," *J.Opt.Soc. Am. B*, 10, 283-95(1993).

[3] C.J.Reilly, W.J.chappell, J.W.halloran, and L.P.katehi, "High-Frequency Electromagnetic Bandgap Structures Via Indirect Solid Freeform Fabrication," J.Am.Ceram.Soc., 87, 1445-53 (2004).

[4] J.S.Foresi, P.R.Villeneuve, J.Ferrera, E.R.Thoen, G.Steinmeyer, S.Fan, J.D.Joannopoulos, L.C.Kimerling, H,I,Smith, and E.P.Ippen, "Photonic Bandgap Microcavities in Optical Waveguides," *Nature*, 390, 143-5 (1997).

[5] Z.Liu, S.Kirihara, Y.Miyamoto, and D.Zhang, "Microwave Absorption in Photonic Crystals Composed of SiC/Resin with a Diamond Structure," *J.Am.Ceram.Soc.*, 89, 2492-95 (2006).

[6] M.W.Takeda, S.Kirihara, Y.Miyamoto, K.Sakoda, and K.Honda, "Localization of Electromagnetic Waves in Three-dimensional Fractal Cavities," *Phys. Rev. Lett.*, 92, 093902-1 (2004).

[7] K.Sakoda, S.Kirihara, Y.Miyamoto, M.W.Takeda, and K.Honda, "Light Scattering and Transmission Spectra of the Menger Sponge Fractal," *Appl. Phys.*, B81, 321-24 (2005).

[8] E.Semouchkina, Y.Miyamoto, S.Kirihara, G.Semouchkin, and M.Lanagan, "Simulation and Experimental Study of electromagnetic Wave Localization in 3D Dielectric Fractal Structures," *Prof. 36th European Microwave Conf.*, (Manchester, Sept. 2006), pp.776-79 (2006).

[9] C.Sun, N.Fang, D.M.Wu, and X.Zhang, "Projection Micro-stereolithography Using Digital Micro-Mirror Dynamic Mask," *Sensors Actuat. A*, 121, 113-20 (2005).

[10] http://www.d-mec.co.jp/

[11] W.Chen, S.Kirihara, and Y.Miyamoto, "Fabrication of Three-Dimensional Micro Photonic Crystals of Resin-Incorporating TiO_2 Particles and Their Terahertz Wave Properties,", *J.Am.Ceram.Soc.*, (2007) in printing.

[12] W.Chen, S.Kirihara, and Y.Miyamoto, "Freeform Fabrication of Photonic Crystals with

3-Dimensional Diamond Structure by Micro-Stereolithography," *Proc. The 2ⁿᵈ Int. Conf. on the Characterization and Control of Interfaces (ICCCI 2006)* (Kurashiki, September, 2006), *Cermam. Trans.* (2007) in printing.

[13] C.Sirtori, "Bridge for the Terahertz Gap," *Nature*, 417, 132-33 (2002).

[14] P.F. Jacobs, Stereolithography and Other RP&M Technologies, Society of Manufacturing Engineering, Dearborn (1996).

[15] J.W.Haus, "A brief review of theoretical results for photonic band structures," *J.Mod. Opt.*, 41, 195-207(1994).

[16] K.M.Ho, C.T.Chan, and C.M.Soukoulis, "Existence of a Photonic Gap in Periodic Dielectric Structures," *Phys. Rev. Lett.*, 65, 3152-55(1990).

[17] Y.Miyamoto, S.Kirihara, and M.W.Takeda, "Localization of Electromagnetic Wave in 3D Periodic and Fractal Structures," *Chem. Lett.*, 35, 342-47 (2006).

FABRICATION OF NEW DIELECTRIC FRACTAL STRUCTURES AND LOCALIZATION OF ELECTROMAGNETIC WAVE

Y. Nakahata, S. Kirihara and Y. Miyamoto

Smart Processing Research Center, Joining and Welding Research
Center, Osaka University, Ibaraki, Osaka 567-0047, Japan

ABSTRACT

New dielectric fractal structures of Menger-sponge cubes were designed by 3D-CAD software and fabricated by stereolithography. The fractal dimension is 2.50 which is different from 2.73 for a normal Menger-sponge. The stage 1, 2, and 3 sponge structures were fabricated. They were composed of 10 vol.% TiO_2-SiO_2 /epoxy with the edge length of 64 mm. The strong localization of electromagnetic waves was observed at about 22 GHz and 27 GHz in the stage 2 and 3 sponges, respectively. It is thought that the localization of electromagnetic wave appears due to the resonance enhanced with the self-similar structure of Menger-sponge.

INTRODUCTION

Fractal is defined as the self-similar structure. Menger-sponge is a typical 3D fractal structure as illustrated in Fig. 1.[1] It is designed by dividing a cubic initiator (a) into 3×3×3 identical smaller cubes, and extracting seven cubes at the face and body center positions, forming the stage 1 Menger-sponge called generator (b). By repeating the same process in remaining twenty smaller cubes, the stage 2 (c) and stage 3 (d) Menger-sponges are obtained. The fractal dimension as a measure of the complexity of fractal structure is calculated as $D = log N / log S$, where N is number of the remaining self-similar units when the size of the initial unit reduces to $1/S$ by a dividing and extracting operation. In case of normal Menger-sponge, the fractal dimension is about 2.73.

Recently, we have succeeded in fabrication of Menger-sponge structure made of dielectric material using a CAD/CAM stereolithography, and found the strong localization of electromagnetic waves with the specific wave length in the structure. We named the fractal structure with the localization function as photonic fractal.[2-4] In order to control the localized frequency it is useful to change the size, material and stage number of the sponge. In this study, new Menger-sponge structures which are different from the normal Menger-sponge were designed by changing the division and extraction numbers and fabricated for different stage structures from 1 to 3. Periodic structures with the same square holes as those of new sponges were also fabricated in order to analyze the localization behavior.

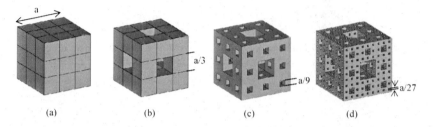

Fig. 1 Images of normal Menger-sponge structures. (a) stage 0, (b) stage 1, (c) stage 2, (d) stage 3.

EXPERIMENTAL

Figure 2 illustrates model structures of new Menger-sponges with four different stages. They were designed by dividing a cubic initiator (a) into 4×4×4 identical smaller cubes and extracting thirty-two cubes at the face and body center positions. (b) is stage 1. By repeating the same process in remaining thirty-two smaller cubes, stage 2 (c) and stage 3 (d) structures are obtained. We called these new fractal structures as (4,2) Menger-sponge. Fractal dimension of (4,2) Menger-sponge is 2.50. In case of the stage 3 (4,2) Menger-sponge, three kinds square holes of $a/2$, $a/8$, $a/32$ in edge length exist where a is the cube edge length. Figure 3 shows model structures of two periodic structures with two different holes of $a/8$ and $a/32$. These holes are arranged in a simple cubic symmetry.

The fractal and the periodic structures were designed on a computer by 3D-CAD software (Toyota Caelum Co. Ltd., Think Design Ver. 9.0). The designed CAD models were converted into STL files and sliced into thin sections. The sliced CAD data was transferred into a stereolithographic machine (D-MEC Co. Ltd. Japan, SCS-300P). The complex structures were formed layer by layer. Each layer was solidified by scanning a laser beam on the surface of liquid epoxy photo-polymer with TiO_2-SiO_2 particles dispersion. The averaged particle diameter is 10 μm, and the volume fraction 10 %. The ultraviolet laser beam of 350 nm in wavelength and 100 μm spot diameter was scanned with 90 mm/s in speed. The thickness of each layer was about 100 μm.

The electromagnetic wave attenuation of the transmission through samples was measured by using two horn antennas and a network analyzer (Agilent Technologies, E8364B) as illustrated in Fig. 4-(a). The spectra were obtained by using the time domain method in order to cut off multiple reflections. The localized waves in a fractal structure scatter to all directions besides being absorbed due to the dielectric loss of material. In order to detect the scattered peak of localized wave, 90° scattering was measured in an experimented setup shown in Fig. 4-(b). [6] Spatial distribution profiles of electric field at the center horizontal plane of fractal and periodic structures through an aluminum aperture were measured by using a monopole antenna in the inner and outer space of the sponges. In order to move a mono-pole antenna at constant intervals automatically, a stepping motor controller (Suruga Seiki Co. Ltd, D220) was used.

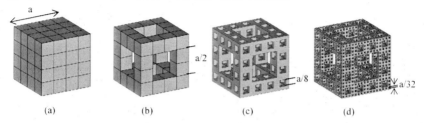

Fig. 2 Images of new sponge structures. (a) stage 0, (b) stage 1, (c) stage 2, (d) stage 3.

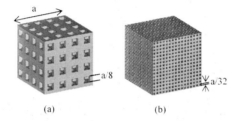

Fig. 3 Images of periodic structures. (a) periodic 1, (b) periodic 2.

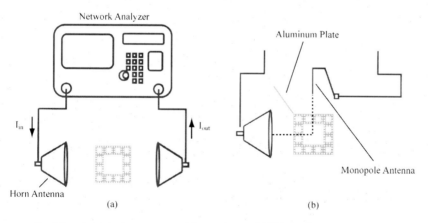

Fig. 4 Experimental setups to measure (a) transmittance, (b) 90°scattering wave spectra.

RESULTS AND DISCUSSION

Figure 5 shows a CAD model of the stage 3 (4,2) Menger-sponge and a fabricated sample composed of TiO₂-SiO₂ particles dispersed epoxy resin. We could precisely fabricate fractal structures and periodic structures based on CAD data. The side length is 64 mm and three kinds of square holes are 2 mm, 8 mm, and 32 mm in edge length. The dielectric constant of the component material is 8.7.

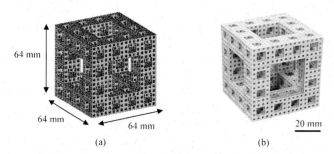

(a)   (b)

Fig. 5 (a) a CAD model with stage 3 (4,2) Menger-sponge structure, (b) a fabricated sample with epoxy dispersed TiO₂-SiO₂ particles.

Figure 6 shows the transmission spectra of the stage 1, 2, and 3 sponges and two periodic structures. The sharp attenuations of transmittance were observed at 22.1 GHz for the stage 2 sponge and 27.0 GHz for the stage 3. In addition, the attenuation peak of stage 3 was sharper than that of stage 2. In contrast, such sharp attenuation of transmittance was not observed for the stage 1 sponge and two periodic structures in the measured frequency range from 15 GHz to 30 GHz. A wide attenuation band of transmittance from about 23.0 GHz to 29.0 GHz was observed in the periodic 2 structure. The periodic 1 and 2 structures of hole lattices can totally reflect electromagnetic waves based on Bragg's law. [7] Therefore, the forbidden band called photonic band gap is formed in a certain frequency range. In case of the periodic 1 structure, the photonic band gap will be formed in a lower frequency against the measured range.

The intensity profiles of the electric field for the stage 2 sponge at 22.1 GHz and the stage 3 at 27.0 GHz are shown in Fig. 7. The intensity of transmitted wave is weak in both fractals. The electric field intensity around the stage 3 sponge is weaker than that around the stage 2. These results support the localization in the stage 2 and 3 sponges at the corresponding frequencies. The localization function of the stage 3 sponge was stronger than that of the stage 2.

Figure 8 shows the 90°scattering wave spectra for the stage 1, 2 and 3 sponges. Highly scattered peaks were observed at 22.1 GHz for the stage 2 sponge and 27.0 GHz for the stage 3, corresponding to the frequencies of attenuation dips in transmission spectra. In contrast, such a sharp dip was not observed for the stage 1 sponge. Quality factors Q of the localized modes for the stage 2 and 3 sponges were calculated about 1200 and 1400 by using the equation of $Q=f/\Delta f$, respectively, where f is the mid frequency of the scattered peak and Δf is the peak width at the half-maximum.

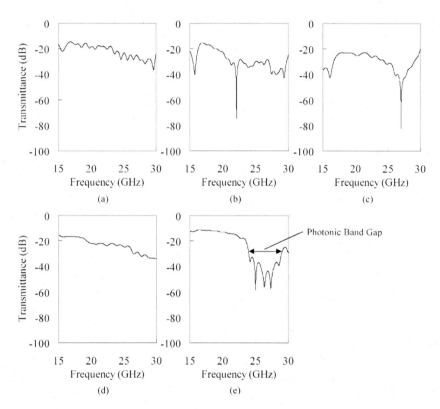

Fig. 6 Transmission spectra of (a) stage 1, (b) stage 2, and (c) stage 3 sponge structures, (d) periodic 1, (e) periodic 2 structures.

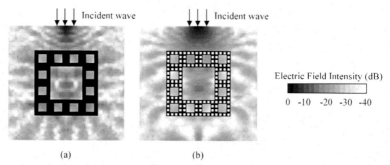

Fig. 7 Intensity profiles of the electric field around the (b) stage 2 sponge at 22.1GHz, and (c) stage 3 at 27GHz.

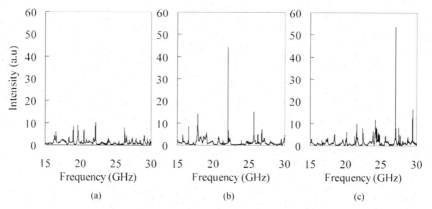

Fig. 8 Spectra of the 90°wave scattering which are normalized with the intensity in free space, (a) stage 1, (b) stage 2, and (c) stage 3 sponges.

The wavelengths of localization mode in a dielectric Menger-sponge can be predicted using the empirical equation which was derived in our previous studies, [4]

$$\lambda = \frac{r'a\sqrt{\varepsilon_{eff}} \cdot n}{S^{2^{r}-1}}$$

where λ is the wavelength of the localized mode in air, r is the remaining number of divided side edges after the subtraction process, n is subtracted number of divided side edges, a is the side length of the Menger-sponge cube, ℓ is the order number of the localized modes, and S is the division number of a side edge of the Menger-sponge. ε_{eff} is the volume averaged effective dielectric constant of the Menger-sponge which is calculated by the mixing rule of ε_{eff} $=V_f\varepsilon_A+(1-V_f)\varepsilon_B$, where ε_A and ε_B are dielectric constants of air and object material, respectively. V_f is the volume fraction of the material of a Menger-sponge which is expressed by $f=(N/S^n)^m$, where n is the dimension number of the structure $(0,1,2,3\ldots)$ and m is the stage number of the Menger-sponge.

We applied this empirical equation to $(4,2)$ Menger-sponge. In case of $(4,2)$ Menger-sponge, $S = 4$ and $n = 2$. Table 1 compares the calculated frequencies with measured ones. The first order modes with lower frequencies were out of measurable range. In case of the stage 2 sponge, the effective dielectric constant was 2.93 and the calculated second mode frequency was 21.9 GHz. The measured frequency at 22.1 GHz was very close to the calculated frequency. In case of the stage 3 sponge, the effective dielectric constant was 1.96 and the calculated second mode frequency was 26.8 GHz. The measured frequency at 27.0 GHz was also very close to the calculated frequency. These results suggest that the empirical equation which was obtained by studies of the normal Menger-sponge can be applied to the new $(4,2)$ Menger-sponges.

This equation suggests that the localized wave with 8 wavelengths resonate as the second order mode in a $(4,2)$ Menger-sponge structure with effective dielectric constant ε_{eff}. In case of the stage 1 sponge, the effective dielectric constant was 4.85 and the calculated second mode frequency was 17.0 GHz. However, no localization was observed as seen in Fig. 6-(a) and Fig. 8-(a). These results may suggest that the localized mode can exist in the stage 1 sponge, but the electromagnetic wave can not resonate strongly because of the simple open structure. It is considered that the self-similar structures of the stage 2 and 3 sponges enhance the resonance of the localized mode.

Table 1 Calculated and measured localized frequencies.

	ε_{eff}	Calculated (GHz)		Measured Frequency (GHz)
		$\ell = 1$	$\ell = 2$	
stage 1	4.85	2.13	17.0	-
stage 2	2.93	2.74	21.9	22.1
stage 3	1.96	3.35	26.8	27.0

CONCULUSION

We fabricated new dielectric fractal structures named (4,2) Menger-sponge and related periodic structures of square holes composed of epoxy with 10 vol.% TiO_2-SiO_2 particles dispersion using stereolithography. The localization of electromagnetic wave was observed for the stage 2 and 3 sponges. The localization function of the stage 3 sponge was stronger than that of the stage 2. However, no localization was observed in the stage 1 sponge and periodic structures. The calculated frequencies of localization using the empirical equation derived for normal Menger-sponges showed good agreements with the measured frequencies. These results suggested that the empirical equation can be used to the new (4,2) Menger-sponges, and the self-similar structure can enhance the resonance of the localized mode.

ACKNOWLEDGEMENTS

This work is supported by the Grant for the 21st Century's COE Program "Center of Excellence for Advanced Structural and Functional Materials Design", and by the Grant-in-Aid for Scientific Research (S) No. 171067010, both under the auspices of the Ministry of Education, Culture, Sports, Science and Technology, Japan.

REFERENCES

[1] B. B. Mandelbrot, "The Fractal Geometry of Nature," (W. H. Freeman & Company, San Francisco, CA, 1982).

[2] M. W. Takeda, S. Kirihara, Y. Miyamoto, K. Sakoda, and K. Honda, "Localization of electromagnetic waves in three-dimensional fractal cavities," *Phys. Rev. Lett.* **92**, 093902-1 (2004).

[3] Y. Miyamoto, S. Kirihara, S. Kanehira, M. W. Takeda, K. Honda, and K. Sakoda, "Smart processing development of photonic crystals and fractals," *Int. J. Appl. Ceram. Technol.* **1**, 40-48 (2004).

[4] Y. Miyamoto, S. Kirihara, M.W. Takeda, K. Honda and K. Sakoda, "A New functional Material: Photonic Fractal," *Functionally Graded Mater.* ⊔, 77-83 (2004).

[5] S. Kirihara, M. W. Takeda, K. Sakoda, K. Honda, and Y. Miyamoto, "Strong localization of microwave in photonic fractals with Menger-sponge structure," *Int. J. Appl. Ceram. Technol.* **26**, 1861-1864 (2005).

[6] K. Sakoda. "90 degree light scattering by the Menger sponge fractal," *Opt. Express.* **13**, 9585-9597 (2005).

[7] M. W. Takeda, Y. Doi, K, Inoue, Joseph W. haus, Zhenyu Yuan. "A simple-cubic photonic lattice in silicon," *Appl. Phys. Lett.* **70**, 2966–2968.

MICRO-FABRICATION AND TERAHERTZ WAVE PROPERTIES OF ALUMINA PHOTONIC CRYSTALS WITH DIAMOND STRUCTURE

H. Kanaoka, S. Kirihara and Y. Miyamoto
Smart Processing Research Center, Joining and Welding Research Center,
Osaka University, Ibaraki, Osaka 567-0047, Japan

ABSTRACT

Fabrication and terahertz wave properties of alumina micro photonic crystals with a diamond structure were investigated. Three-dimensional diamond structure was designed on a computer by a 3D-CAD software. The designed lattice constant was 500 μm. The structure consisted of 8×8×4 unit cells. Acrylic diamond structures with alumina dispersion at 40 vol. % were formed by using micro-stereolithography. After dewaxing acrylic resin at 600°C in air, they were sintered at 1500°C. The linear shrinkage ratio was about 25 %. The relative density reached 97.5 %. The electromagnetic wave properties were measured by terahertz time-domain spectroscopy. A wide complete photonic band gap was observed at the frequency range from 0.40 THz to 0.47 THz, which was in a good agreement with the simulation result calculated by plane wave expansion method.

INTRODUCTION

Periodically arranged structures of dielectric media are called photonic crystals[1]. They have photonic band gaps in which no electromagnetic wave can propagate. If they have a structural defect, localized modes appear in the band gap. Such localization of electromagnetic waves can be applied to resonators, waveguides, and antennas. Three-dimensional photonic crystals with a diamond structure have been regarded as the ideal crystal since they can prohibit the propagation of electromagnetic waves for any directions in the band gap[2]. However, due to the complex structure, it was difficult to fabricate them. Our group has succeeded in fabricating millimeter-order diamond structures by a CAD/CAM stereolithography and demonstrated that they have a complete photonic band gap in microwave region[3].

Here, we present a new technique to fabricate micrometer-order diamond structures and their terahertz wave properties. In recent years, terahertz waves have received extensive attention and investigation since they have a lot of interesting and applicable features in many fields such as material, communication, medicine, and biology[4]. Dense alumina micro diamond structures were fabricated by dewaxing and sintering of the green bodies composed of nano-sized alumina particles dispersed in acrylic resin which were formed by micro-stereolithography. Terahertz waves are absorbed by water vapor in air[5], thus such low-loss ceramics as alumina is suitable for propagation and control of terahertz waves.

DESIGN OF DIAMOND LATTICE

Electromagnetic band diagrams of diamond structures were calculated by plane wave expansion method to determine their geometric parameters. The dielectric constant of the lattice used in calculation was 10 for alumina. Figure 1(a), (b), and (c) show a unit cell of the diamond structure, the definition of the aspect ratio, and the calculated complete band gap width as a function of the aspect ratio, respectively. According to Fig. 1(c), the band gap becomes the widest when the aspect ratio is 2.0. The wider the band gap, the easier it is to localize the electromagnetic waves when a defect is introduced. However, it is difficult to fabricate thin lattice. When the aspect ratio is 1.5, the lattice thickens and the band gap width is approximately 84 % as much as that of 2.0. Thus the aspect ratio of the diamond structure was designed to be 1.5. The lattice constant was 500 μm. The whole structure was 4×4×2 mm³ in size, consisting of 8×8×4 unit cells.

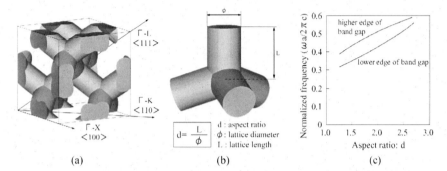

Fig. 1 (a) A CAD image of a unit cell of diamond structure, (b) Definition of aspect ratio, and (c) Band gap width as a function of aspect ratio.

EXPERIMENTAL PROCEDURE

Three-dimensional diamond structure was designed by using a 3D-CAD program (Toyota Caelum Co. Ltd., thinkdesign ver. 7.0). The CAD data was converted into a STL file of a rapid prototyping format. After the slicing process of 3D model into a series of two-dimensional cross-sectional data, they were transferred to micro-stereolithography equipment (D-MEC Co. Ltd., Japan, SI-C1000). There are several methods to fabricate micrometer-order structures[6,7]. In our system, photo sensitive acrylic resin dispersed with alumina particles of 170nm in diameter at 40 vol. % was fed over the substrate from a dispenser nozzle. The high viscosity ceramic / resin paste was supplied with pressured air. Then it was spread uniformly by moving a knife edge. Two-dimensional pattern of 10 μm in thickness was formed by exposing visible laser of 405 nm in wavelength on the resin surface. The high resolution has been achieved by using a Digital Micromirror Device (DMD) and an objective lens. Figure 2 shows a schematic of the micro-stereolitography system. The DMD is an optical element assembled by mirrors of 14 μm in edge length. Each tiny mirror can be controlled according to the two-dimensional cross sectional data by a computer. Three-dimensional structure was built by stacking these patterns layer by layer.

In order to avoid deformation and cracking during dewaxing, careful investigation for the dewaxing process is required[8]. The green bodies with diamond structure were heated at various temperatures from 100 °C to 600 °C while the heating rate was fixed at 1.0 °C /min. The dewaxing process was observed in respect to the weight and color changes. The sintering temperature and the heating rate were fixed at 1500 °C and 8 °C/min, respectively. The density of the sintered sample was measured by Archimedes method. The microstructure was observed by scanning electron microscopy (SEM). The transmittance and the phase shift of incident terahertz waves were measured by using terahertz time-domain spectroscope (Advanced Infrared Spectroscopy Co. Ltd., Japan, J-Spec2001 spc/ou).

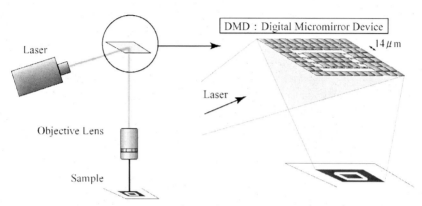

Fig. 2 A schematic illustration of micro-stereolithography.

RESULTS AND DISCUSSION

Figure 3 shows an alumina-dispersed resin sample fabricated by micro-stereolithography. The spatial resolution was approximately 0.5 %. The weight and color changes as a function of temperature are shown in Fig. 4. The sample color changed into black at 400 °C due to carbonizing of resin. It became white at 600 °C and the weight decrease saturated. The dewaxing process is considered to start at 200 °C and complete at 600 °C. Thus the dewaxing temperature was optimized to be 600 °C. By the dewaxing and sintering process, full-ceramic diamond structures were successfully obtained. Figure 5 (a), (b), and (c) show the sintered samples along different crystal directions. The deformation and cracking were not observed. The linear shrinkage on the horizontal axis was 23.8 % and that on the vertical axis was 24.6 %. The relative density reached 97.5 %. Dense alumina microstructure was formed as shown in Fig. 5 (d). The average grain size was approximately 2 µm.

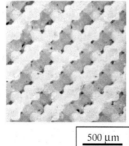

500 µm

Fig. 3 An acrylic diamond lattice with alumina dispersion formed by micro-stereolithography.

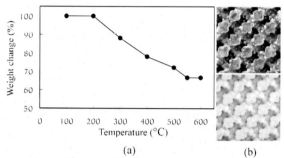

(a)   (b)

Fig. 4 (a) The weight change as a function of temperature and (b) the lattice color when the dewaxing temperature is 400 °C (upper) and 600 °C (lower), respectively.

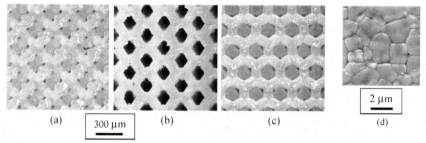

(a)   300 μm   (b)   (c)   (d)

Fig. 5 Sintered alumina photonic crystals with a diamond structure. (a) (100) plane, (b) (110), and (c) (111). (d) microstructure of alumina lattice.

Figure 6 (a), (b), and (c) show the transmission intensity and the phase shift spectra along Γ-X <100>, Γ-K <110>, and Γ-L <111> directions, respectively. A wide common band gap was observed in every direction at the frequency range from 0.40 THz to 0.47 THz. When a gap is formed, there are two types of the standing wave modes with the wavelength corresponding to periodicity at the frequencies of the each band edges. The lower frequency mode concentrates the wave energy in the dielectric region, whereas the higher frequency mode concentrates in the air region. Thus the difference between the phase of these modes must be π across the gap[9]. The gradient of the phase shift became large at the band gap edges, indicating that the group velocity of electromagnetic waves was very small. The band gap frequencies were compared with the calculation by plane wave expansion method (Fig. 7). The measured opaque region corresponded to the calculation. Furthermore, according to the photonic band diagram, it was confirmed that a complete photonic band gap opened between 0.40 THz and 0.47 THz. These three-dimensional photonic band gap structures can be applied to control terahertz waves.

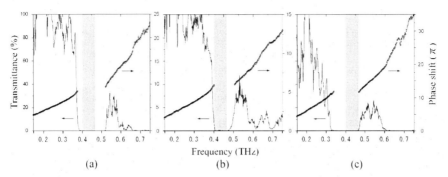

Fig. 6 Frequency dependences of the transmittance and phase shift spectra along (a) Γ-X <100>, (b) Γ-K <110>, and (c) Γ-L <111> directions. The frequency ranges with gray color indicate the common band gap.

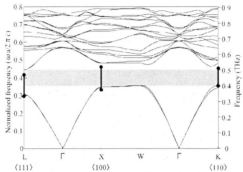

Fig. 7 Photonic band diagram calculated by plane wave expansion method. Plots are measured photonic band gap frequencies. The dielectric constant of the lattice was 10. The frequency range with gray color indicates the perfect band gap in common for all crystal directions.

CONCLUSIONS

We have formed three-dimensional micro photonic crystals with a diamond structure composed of alumina dispersed acrylic resin at 40 vol. % by micro-stereolithography. By the careful optimization of process parameters regarding dewaxing and sintering, we have succeeded in fabricating dense alumina micro diamond structures. The sintered photonic crystal of alumina formed a complete photonic band gap at the terahertz region from 0.40 THz to 0.47 THz. The measured band gap showed a good agreement with the calculation by plane wave expansion method.

ACKNOWLEDGEMENTS

This work is supported by the Grant for the 21st Century's COE Program "Center of Excellence for Advanced Structural and Functional Materials Design", and by the Grant-in-Aid for Scientific Research (S) No. 171067010, both under the auspices of the Ministry of Education, Culture, Sports, Science and Technology, Japan.

REFERENCES

[1] E. Yablonovitch, "Inhabited Spontaneous Emission in Solid-State Physics and Electronics," *Phys. Rev. Lett.*, **58**, 2059-62 (1987).

[2] K. M. Ho, C. T. Chan, and C.M. Soukoulis, "Existence of a Photonic Gap in Periodic Dielectric Structures," *Phys. Rev. Lett.*, **65**, 3152-55 (1990).

[3] S. Kirihara, Y. Miyamoto, K. Takenaga, M. W. Takeda, and K. Kajiyama, "Fabrication of Electromagnetic Crystals with a Complete Diamond Structure by Stereolithography," *Solid State Communications*, **121**, 435-39 (2002).

[4] R. M. Woodward, V. P. Wallace, D. D. Arnone, E. H. Linfield, and M. Pepper, "Terahertz Pulsed Imaging of Skin Cancer in the Time and Frequency Domain," *Journal of Biological Physics*, **29**, 257-61 (2003).

[5] M. Van Exter, Ch. Fattinger, and D. Grischkowsky, "Terahertz Time-Domain Spectroscopy of Water Vapor," *Optics Letters*, **14**, 1128-30 (1989).

[6] X. Zhang, X. N. Jiang, and C. Sun, "Micro-Stereolithography of Polymeric and Ceramic Microstructures," *Sensors and Actuators*, **77**, 149-56 (1999).

[7] J. A. Lewis and G. M. Gratson, "Direct Writing in Three Dimensions," *Materials Today*, **7**, 32-39 (2004).

[8] S. Kanehira, S. Kirihara, and Y. Miyamoto, "Fabrication of TiO_2-SiO_2 Photonic Crystals with Diamond Structure," *J. Am. Ceram. Soc.* **88**, 1461-64 (2005)

[9] T. Aoki, M. W. Takeda, J. W. Haus, Z. Yuan, M. Tani, K. Sakai, N. Kawai and K. Inoue, "Terahertz Time-Domain Study of a Pseudo-Simple-Cubic Photonic Lattice," *Phys.Rev. B*, **64**, 045106-1-5 (2001).

TEXTURE DEVELOPMENT OF $Bi_4Ti_3O_{12}$ THICK FILM PROMOTED BY ANISOTROPIC SHRINKAGE

Y. Kinemuchi, P.H. Xiang, H. Kaga and K. Watari
Advanced Industrial Science and Technology (AIST)
2266-98 Anagahora, Shimoshidami, Moriyama,
Nagoya 463-8560, Japan

ABSTRACT

The progress of texturing during film processing has been observed for $Bi_4Ti_3O_{12}$ thick film. Initially, the texture of screen-printed film was found to be at random orientation, which was attributed to the equiaxed particle shape of synthesized powders. Subsequent heating resulted in a reduction of thickness as densification progressed. On the contrary, in-plane shrinkage did not occur due to constraint imposed by the substrate, and this anisotropic shrinkage was expected to cause shear deformation of the film. The heating process also promoted the morphological change of the particles from equiaxial to platy. Owing to both effects during the heating, the orientation degree of c-axis increased in proportion to the thickness reduction, as analyzed by the March-Dollase function. When the film was fully densified, a Lotgering factor of 0.5 was achieved. Prolonged heat treatment at 1050 °C further enhanced the degree of orientation due to grain growth, in which a linear correlation between the Lotgering factor and grain size was found. Finally, highly c-axis oriented thick film with a Lotgering factor of 0.98 was successfully obtained.

INTRODUCTION

Texturing is an effective way of improving the properties of materials. In ceramic bulk materials, the texturing is usually achieved by using mechanical deformation, preferential grain growth from seed crystals or applying magnetic field. For instance, hot forging results in textured ceramics due to a shear force generated during plastic deformation, which induces recrystallization and/or grain rotation.[1] On the other hand, templated grain growth is based on the preferential grain growth from seed crystals aligned one or two dimensionally by a shear force during forming.[2] It is also possible to achieve particle alignment by applying a magnetic field when the particle possesses anisotropy in magnetic susceptibility.

The anisotropy of ferroelectric properties in bismuth titanate, $Bi_4Ti_3O_{12}$ (BiT), is fairly large, for instance, there is a more than ten-fold difference in the component of polarization between the a and c axes[3,4], which is attributed to its layered structure composed of two $BiTiO_3$ unit cells of hypothetical perovskite structure interleaved with BiO_2 layers[5]. Hence, the texturing of this material enhances the properties compared to the randomly oriented one, which was demonstrated by hot forging[6], templated grain growth[7] and magnetic field alignment[8] for bulk materials. Furthermore, texturing of thin films has been successfully carried out by using epitaxial growth[9] or sol-gel coating of a well-designed organic precursors[10].

This paper focuses on the texturing of BiT thick film. Thick film technology is based on sintering of powdered materials on substrates, which is similar to that of bulk materials; however, the influence of the substrate is significant. One of the distinctive features of thick film densification is that there is little shrinkage in the in-plane direction during sintering. This constraint is enough to cause shear deformation, which may bring particle rotation. To demonstrate the nature of texture evolution in thick film, equiaxed particles as starting material were prepared by co-precipitation and its paste was coated by screen-printing on substrates. The effect of subsequent heating on the texturing is reported here.

EXPERIMENTAL

Coprecipitation method was used to synthesize BiT particles. Bismuth nitrate $(Bi(NO_3)_3 \cdot 5H_2O$, 99.5%, Wako Pure Chemical Industries, Ltd., Japan.) was initially dissolved in HNO_3 solution (Analytically, Wako Pure Chemical Industries, Ltd., Japan) at pH>3 to produce a clear solution, and then titanium tetra-n-butoxide $(Ti(O-n-C_4H_9)_4$, 99%, High Purity Chemical, Japan) ethanol solution was slowly added with continuous stirring. Subsequently, ammonia $(NH_3 \cdot H_2O$, Analytically, Wako Pure Chemical Industries, Ltd., Japan) was added dropwise to the above homogeneous solution under vigorous stirring to produce white precipitate at pH>8. Finally, the obtained precipitate was thoroughly washed with dilute ammonia and ethanol, dried, and calcined at 600 °C for 1 h. The as calcined powder was composed of single phase of $Bi_4Ti_3O_{12}$ with a primary size of ca. 100 nm. Its morphology was almost equiaxial in shape. The details of this powder such as powder morphology, constitutive phase and sintering behavior are described elsewhere[11].

The powder was then mixed with α-terpineol and methyl cellulose, which was subsequently used as a paste for screen printing. The solid loading of this paste was adjusted to be 25%. The amount of cellulose was 10 vol% of the powder volume. The paste was coated on polished alumina substrates by screen printing. After printing, the films were dried at 50 °C. The film thickness after drying was found to be 20 – 50 μm.

Sintering was carried out in the temperature range of 800 – 1050 °C, in a lidded crucible of alumina. To compensate any Bi loss by evaporation during sintering, BiT powder was also settled in the crucible.

The orientation degree of the film was conventionally evaluated by the Lotgering factor[12] based on the diffraction intensities of (004), (006), (008), (111) and (-117) peaks, where reference intensities for the random orientation case were obtained from the JCPDS file (#80-2143). The pole-figure profiles were observed using the Schulz reflection method. The profiles were analyzed based on the March-Dollase function[13] after eliminating background and defocusing effects[14].

The density of films was measured from the film weight divided by the film volume. The surface profile of the films measured by a confocal LASER scanning microscope was used to obtain the film volume. Here, the scanning was carried out on entire surface of the films.

RESULTS AND DISCUSSION

Figure 1 shows the film density as a function of film weight divided by thickness. A linear relation between them can be seen. Because the density was derived from the film weight divided by film volume, the linear relationship in Fig. 1 indicates that the film area was constant during heating. In fact, the inverse of the slope well corresponded to the actual film area coated by screen printing. This means that densification of thick film mainly proceeded from thickness reduction without in-plane shrinkage. Taking into account the relative density of green body of 0.5, the thickness reduction reached half of the original thickness when full densification was achieved, and so the thickness reduction ratio during heating ranged between 0 and 0.5.

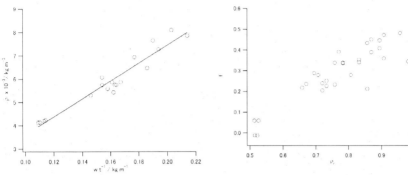

Fig. 1 Film density (ρ) as a function of film weight (w) normalized by thickness (d).

Fig. 2 The Lotgering factor (f) as a function of relative density (ρ_r).

Figure 2 shows the relation between the Lotgering factor (c-axis orientation) and density. As expected, the Lotgering factor increased with increase in density. For the set of these data, the holding temperature was varied in the range of 400 – 1050 °C and the temperature was maintained for 1 h. Because of the equiaxial shape of the starting powder, the Lotgering factor after binder-burnout (i.e. Lotgering factor at ρ_r of 0.5) was almost zero, indicating that screen printing had no effect on texture.

One of the possible reasons for the increase in the Lotgering factor is particle rotation during shrinkage. As shown in Fig. 1, increase in density led to a reduction of thickness without shrinkage in the in-plane direction. Because of this characteristic, the situation of particles was identical with sintering under a shear stress field. Although the starting powder had an equiaxial shape, the shape of the particles changed into platelet morphology during sintering as shown in Fig. 3. This anisotropic shape of particles enabled the particles to orient under shear stress.

The progress of texturing was also confirmed by pole-figure profile as shown in Fig. 4. Here, the distribution of (006) diffraction intensity was monitored; where ϕ is the angle of tilt from the plane normal to the substrate surface. It is clear that the c-axis orientation gradually emerged in the out-of-plane direction ($\phi=0$) as the film density increased. In order to analyze the progress of texturing qualitatively, these profiles were fitted to the March-Dollase function. This distribution function is as follows:

$$\left(r^2\cos^2\phi - r^{-1}\sin^2\phi\right)^{-\frac{2}{3}}$$

Eq.1

This function was derived for the development of preferred orientation as a result of rigid-body rotation of platy or rod-shaped grains upon linear sample deformation. The March coefficient, r, characterizes the strength of the preferred orientation and is related to the amount of sample deformation. For platy crystallites the relation is $r=d/d_0$, where d_0 is the thickness of an original (hypothetical) sample showing uniform pole density and d is the sample thickness after axial extension or compaction.[13] The solid line in Fig. 4 indicates fitted curves using the March-Dollase function, and reasonable fitting to the experimental data is found.

In the curve fitting, the March coefficient was used as the fitting parameter, and the obtained parameter, r, was plotted against the inverse of density as shown in Fig. 5. There is a good linear relation between them. Remembering that the film density linearly increased with

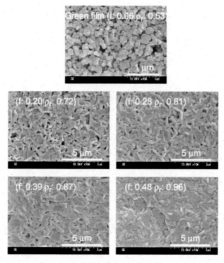

Fig.3 Surface morphology of films.

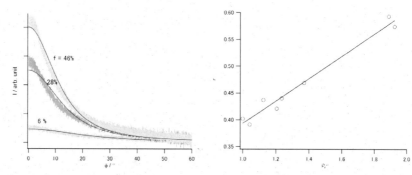

Fig. 4 (006) pole-figure profiles. The lines are best-fitted curves using the March-Dollase function. Here, the Lotgering factor, f, for each sample is also noted.

Fig. 5 Linear relation between March parameter, r, and inverse of relative density (ρ_r).

thickness reduction (see Fig. 1), it is found that the inverse of relative density increased in proportion to film thickness. In addition, the March coefficient also has a linear relation with deformed thickness. Hence, the linear relation between them indicates that the preferential orientation during the film sintering originates in the reduction in the film thickness.

Although the texture was developed by the anisotropic shrinkage of the film, its orientation degree was not high, i.e. f of 0.5 at the theoretical density. Sakka et al. reported that

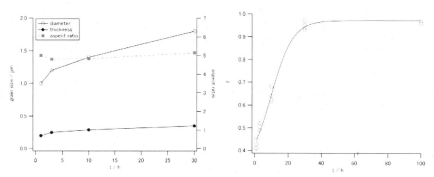

Fig. 6 Grain size progress during heat treatment at 1050 °C. Aspect ratio is that of diameter to thickness.

Fig. 7 Increase of the Lotgering factor, f, during heat treatment at 1050 °C.

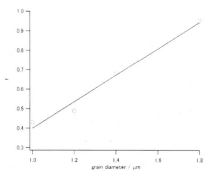

Fig. 8 Progress of orientation degree, the Lotgering factor f, as a function of grain diameter.

the higher degree of orientation could be achieved with recrystallization by grain growth in magnetically textured ceramics.[15] In this situation, larger grains grow at the expense of adjacent smaller grains. Basically, grain growth does not contribute to the development of texture since its evolution is macroscopically random. Nevertheless, preferential grain growth can occur under the circumstance of partly textured ceramics such as magnetically aligned ceramics or doping with seed crystal. That is, the texturing of bulk ceramics gradually changes under the influence of the orientation distribution of large grains, so the preferential orientation of large grains ensures the entire texturing throughout the grain growth period.

In the BiT thick film, prolonged heat treatment increased the grain size while maintaining the aspect ratio of grain dimensions between in-plane and out-of-plane (Fig. 6), while the orientation degree improved as shown in Fig. 7. It was found that the orientation degree was linearly correlated with the grain size, indicating that grain growth effectively contributed to the texture development (Fig. 8). The progress of texturing during heat treatment was also confirmed by the pole figure of (-117). The diffraction from (-117) is the strongest peak in the randomly

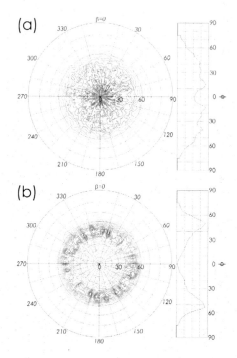

Fig. 9 (-117) pole-figure profiles during heat treatment at 1050 °C. Holding time was 10 and 30 h for (a) and (b), respectively.

oriented BiT, and has a tilt angle of 50.7° from its c-axis. During the progress of texturing, the pole figure showed quite a broad peak as shown in Fig. 9(a). Further progress of texturing resulted in a strong peak near 50° as shown in Fig. 9(b), indicating c-axis orientation of the film. These results support gradual texture development enhanced by grain growth.

CONCLUSION

The progress of texture of Bi$_4$Ti$_3$O$_{12}$ thick film during sintering was observed. As density increased, the Lotgering factor gradually increased and reached 0.5 at full density. The densification during sintering of the thick film mainly proceeded by thickness reduction, leading to shear deformation. This deformation is considered to be the main driving force of the texturing, which was supported by the linear relation between March coefficient and inverse of density. Further increase of orientation degree occurred during prolonged heat treatment after densification. During this period, grain size monotonously increased, and a good correlation between orientation degree and grain size was found, indicating that recrystallization along with grain growth effectively enhanced the texturing.

REFERENCES

[1] A. H. Heuer, D. J. Sellers and W. H. Rhodes, "Hot-working of aluminum oxide: I Primary recrystallization and texture", *J. Am. Ceram. Soc.*, **52**, 468-474 (1969).

[2] M. M. Seabaugh, I. H. Kerscht and G. L. Messing, "Texture development by templated grain growth in liquid-phase-sintered α-alumina", *ibid.*, **80**, 1181-88 (1997).

[3] J. F. Dorrian, R. E. Newnham and D. K. Smith, "Crystal structure of $Bi_4Ti_3O_{12}$", *Ferroelectrics*, **3**, 17-27 (1971).

[4] G. W. Taylor, S. A. Keneman and A. Miller, "Depoling of single-domain bismuth titanate", *ibid.*, **2**, 11-20 (1971).

[5] B. Aurivilius, "Mixed bismuth oxides with layer lattices", *Ark. Kemi.*, **1**, 499-512 (1949).

[6] T. Takenaka and K. Sakata, "Grain-orientation and electrical properties of hot-forged $Bi_4Ti_3O_{12}$ ceramics", *Jpn. J. Appl. Phys.*, **19**, 31-39 (1980).

[7] J. A. Horn, S. C. Zhang, U. Selvaraj, G. L. Messing and S. T. McKinstry, "Templated grain growth of textured bismuth titanate", *J. Am. Ceram. Soc.*, **82**, 921-26 (1999).

[8] A. Makiya, D. Kusano, S. Tanaka, N. Uchida, K. Uematsu, T. Kimura, K. Kitazawa and Y. Doshida, "Particle oriented bismuth titanate ceramics made in high magnetic field", *J. Ceram. Soc. Jpn.*, **111**, 702-704 (2003).

[9] R. Ramesh, K. Luther, B. Wilkens, D. L. Hart, E. Wang and J. M. Tarascon, "Epitaxial growth of ferroelectric bismuth titanate thin films by pulsed laser deposition", *Appl. Phys. Lett.*, **57**, 1505-07 (1990).

[10] H. Gu, D. Bao, S. Wang, D. Gao, A. Kuang and X. Li, "Synthesis and optical properties of highly c-axis oriented $Bi_4Ti_3O_{12}$ thin films by sol-gel processing", *Thin Solid Films*, **283**, 81-83 (1996).

[11] P. H. Xiang, Y. Kinemuchi and K. Watari, "Sintering behaviors of bismuth titanate synthesized by a coprecipitation method", *Mater. Lett.*, **59**, 3590-94 (2005).

[12] F. K. Lotgering, "Topotactical reactions with ferrimagnetic oxides having hexagonal crystal structures – I.", *J. Inorg. Nucl. Chem.*, **9**, 113-23 (1959).

[13] W. A. Dollase, "Correction of intensities for preferred orientation in powder diffractometry: Application of the March model.", *J. Appl. Cryst.*, **19**, 267-272 (1986).

[14] E. M. C. Huijser-Gerits and G. D. Rieck, "Defocusing effects in the reflexion technique for the determination of preferred orientation", *J. Appl. Cryst.*, **7**, 286-290 (1974).

[15] Y. Sakka and T. S. Suzuki, "Textured development of feeble magnetic ceramics by colloidal processing under high magnetic field", *J. Ceram. Soc. Jpn.*, **113**, 26-36 (2005).

ANISOTROPIC PROPERTIES OF Al DOPED ZnO CERAMICS FABRICATED BY THE HIGH MAGNETIC FIELD

Hisashi Kaga[1], Yoshiaki Kinemuchi[1], Koji Watari[1]
[1]National Institute of Advanced Industrial Science and Technology (AIST)
2266-98 Anagahora, Shimo-shidami, Moriyama-ku, Nagoya 463-8560, Japan

Hiromi Nakano[2]
[2]Electron Microscope Laboratory, Ryukoku University
1-5 Seta, Otsu 520-2194, Japan

Satoshi Tanaka[3], Atsushi Makiya[3], Zenji Kato[3], Keizo Uematsu[3]
[3]Department of Materials Science and Technology, Nagaoka University of Technology
1603-1 Kamitomioka, Nagaoka 940-2188, Japan

ABSTRACT

Oriented Al doped ZnO ceramics were fabricated by a high magnetic field via gelcasting technique and their thermoelectric properties and microstructures were investigated. Combination of magnetic field and gelcasting techniques enabled us to align the particles preferentially along the c-axis within a short exposure time in the high magnetic field. The particle orientation was neither degraded nor disturbed by the gelation and subsequent processing, realizing a highly oriented specimen. The electrical conductivity of the c-axis oriented specimen along the ab-plane was almost two times larger than that along the c-axis. On the other hand, the Seebeck coefficients and thermal conductivities exhibited a small anisotropy. A high-resolution transmission electron microscopy observation revealed that an interfacial layer or segregation of Al at grain boundaries was not observed, however, the coincidence site lattices were observed on the highly oriented specimen. This study suggested that highly controlled microstructure and grain boundary were essential for obtaining anisotropy in electrical conductivity.

INTRODUCTION

Texture engineering is an effective way for improving the performance of materials. Crystallographic texturing of polycrystalline ceramics has been produced by a variety of techniques such as hot-forging, templated grain growth (TGG) and reactive templated grain growth (RTGG)[1-5]. However, it is difficult to achieve uniform density and texture for the hot-working methods while there are several restrictions of the starting templates for the seeded grain growth.

In contrast to conventional texture engineering techniques, a high magnetic field method is a promised processing technique for aligning the particle and can be applied any materials; i.e., ceramics, polymers, and metals, if they have an anisotropic property of magnetic susceptibilities[6-9]. In the magnetic field, a particle with an anisotropic magnetic susceptibility is rotated to an angle in order to minimize the system energy, $\Delta E = \Delta \chi V B^2 / 2\mu_o$, where $\Delta \chi$ the anisotropic magnetic susceptibility, V the particle volume of each particle, B the applied magnetic field, and μ_o the permeability in vacuum[10]. In other words, a magnetically anisotropic particle is oriented in the direction of the lowest energy in the magnetic field. For the diamagnetic materials ($|\chi_3| > |\chi_2| > |\chi_1| > 0$), the axis with the largest magnitude of the diamagnetic susceptibility ($|\chi_3|$) is oriented perpendicular to the magnetic field but the direction of this axis is randomly oriented within the plane perpendicular to the magnetic field. As a result, the axis with the smallest diamagnetic susceptibility ($|\chi_1|$) tends to be aligned along the magnetic field.

In the case of ZnO ceramics, the c-axis, which has the largest diamagnetic susceptibility ($|\chi_c| > |\chi_a| > 0$), is oriented randomly within the plane perpendicular to the magnetic field, and thus the a-axis is aligned along the magnetic field. However, polycrystalline ZnO ceramic oriented along the c-axis should have excellent properties[11]. Past study has shown that the ZnO is oriented with the a-axis along the field in a static magnetic field[12].

We have applied a rotational high magnetic field via gelcasting technique to fabricate highly c-axis oriented Al doped ZnO (denoted as Al-ZnO) ceramics and their thermoelectric properties and microstructures were examined.

EXPERIMENTAL
Sample preparation
　　Oriented Al-ZnO ceramics were prepared by the rotational high magnetic field alignment via gelcasting technique. A water-based gelcasting system described in Ref. 13 was used in this study. Specific steps in this process other than in Ref. 13 are described below. Acrylamide (AM) as a monomer and methylenebisacrylamide (MBAM) as a cross-linker were dissolved in distilled water to form premix at a weight ratio as shown in Table I.

Table I. Composition of premix for a gelcasting technique

	AM (mass%)	MBAM (mass%)	Distilled water (ml)
premix	7.4	0.2	100

The mixed powders of ZnO (Hakusui Tech Co., Japan) and Al_2O_3 (Sumitomo Chemical, Japan) having a compositon of $Zn_{0.98}Al_{0.02}O$ with 0.5 mass% dispersant (poly(ammonium acrylate) A-6114, Toagosei Co. Ltd., Japan) were then added in the premix to form a 30 vol% slurry. In order to prepare the aqueous suspension of Al-ZnO, a small amount of dispersing agent was determined by the lowest viscosity of the suspension. The suspension was then degassed in a vacuum desiccator. Before casting into a Teflon mold at room temperature in a 10T magnetic field (TM-10VH10, Toshiba, Japan), the initiator (ammonium persulfate solution, APS) and catalyst (tetramethylethylenediamine, TEMED) were added into the suspension. Gelation in the high magnetic field initiated within 10 min by adjusting the amount of catalyst after casting. The mold was rotated at 30 rpm in the horizontal magnetic field during the gelation. After gelation is finished and when the material is demolded from the container, the gelled body maintained its shape perfectly. The sample was then dried in a series of controlled-humidity drying chamber at room temperature to avoid nonuniform shrinkage due to rapid drying. The relative humidity was kept at 95%. The dry green body was placed in a furnace for removing organic substances. The green body was heated at 600°C for 1 h in air with heating rate of 0.25°C/min. After subjecting to cold isostatic pressing at 196 MPa, the dewaxed green body was sintered at 1400°C for 10 h. A reference specimen was also prepared the same procedure as mentioned above without applying the magnetic field.

Characterization

X-ray diffraction patterns and (002) pole figures were measured with CuKα radiation on polished surfaces of both magnetically processed and reference specimens on an X-ray diffractometer (RINT2550, Rigaku, Japan) equipped with a pole figure goniometer. The pole figure measurements were done by keeping the incident and diffraction angles fixed while azimuthal (β: 0°<β<360°, 2.5° steps) scans were carried out around the normal direction, which was perpendicular to the applied magnetic field at various polar angles (α: 0°<α<75°, 2.5° steps). Texture strength was calculated from March-Dollase function[14] using the results of pole figures after defocusing corrections:

$$f(r,\theta) = \left(r^2 \cos^2\theta + \frac{\sin^2\theta}{r} \right)^{-\frac{3}{2}} \tag{1}$$

where θ is the angle between the texture axis and the scattering vector, r is the orientation parameter. The texture axis is along the magnetic field. The orientation parameter characterizes the width of the texture distribution. For a randomly oriented sample $r = 1$ and for a perfectly oriented sample $r = 0$. The microstructure of sintered specimens was observed by FE-SEM (S-4300, Hitachi, Japan) and HRTEM (3000F, JEOL, Japan). Electrical conductivity was measured using a DC standard four-probe method. Seebeck coefficient was calculated from the thermoelectric voltage and temperature difference between the two ends of the samples. Thermal conductivity was calculated from the product of the thermal diffusivity, specific heat capacity, and density.

RESULTS AND DISCUSSION

The results of XRD patterns of the sintered specimens with and without applying magnetic field are shown in Fig. 1. The planes of analysis are in the directions parallel (//) and perpendicular (⊥) to the applied magnetic field as schematic drawings inseted in Fig. 1.

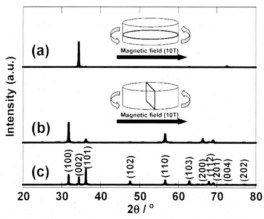

Figure 1. XRD patterns of (a,b)magnetically aligned and (c)reference specimens.

The (//) specimen (Fig. 1a) exhibited high diffraction intensities from (002) and (004) planes, while the diffraction lines of the {00l} planes were hardly observed for the (⊥) specimen (Fig. 1b), realizing a highly c-axis oriented specimen.

Figure 2a shows the texture distribution curves for magnetically aligned and reference specimens with orientation parameters which are also representing the texture strength. The pole figure for the specimen with magnetically aligned specimen is shown in Fig. 2b. The pole figure is presented in gray scale, with the gray scale intensities representing orientation distribution. The orientation parameter calculated from March-Dollase function was 0.23 and 0.94 for magnetically aligned and reference specimens, respectively. The former value represents that about 83% of grains are aligned within 10° from the texture axis while for the latter one it is about 11%. The pole figure clearly shows that the texture is radially symmetric and the (002) pole is very narrow.

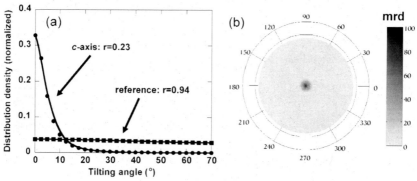

Figure 2. (a) Distribution density curves for magnetically aligned and reference specimens and (b) (002) pole figure of magnetically aligned specimen.

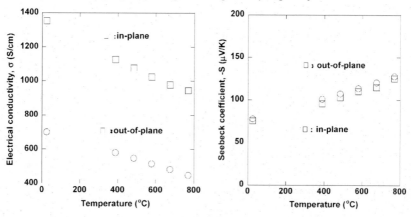

Figure 3. Electrical conductivity and Seebeck coefficient as a function of temperature

The electrical conductivity and Seebeck coefficient as a function of temperature for the magnetically aligned specimen parallel (in-plane) and perpendicular (out-of-plane) to direction of the magnetic field are shown in Fig. 3. In-plane electrical conductivity was almost two times larger than that of out-of-plane over the measured temperatures. On the other hand, the Seebeck coefficients exhibited a small anisotropy regardless of crystallographic orientation. This result suggests that the Seebeck coefficient is rather insensitive to the type of microstructure and thus, the difference in crystallographic direction is small. It should be noted that in-plane and out-of-plane thermal conductivities had similar values.

Figure 4 presents SEM micrographs of the microstructure for the c-axis oriented specimen taken from the directions parallel and perpendicular to the applied magnetic field. The structures differ significantly with directions. The particles elongated along the direction of the applied magnet field (Fig. 4a), while the particles have a near hexagonal shape when observed from the other direction (Fig. 4b). This is due to the increasing interference between adjacent particles in highly oriented particles in the green body. These observations indicate that the particles are oriented and the growth of ab-planes enhances the degree of orientation during the sintering, realizing a highly oriented specimen.

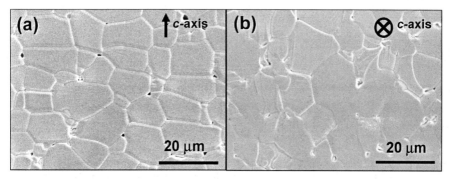

Figure 4. SEM images of magnetically aligned specimen
parallel and perpendicular to applied magnetic field.

Figure 5 shows the HRTEM images of the sintered specimen. The XRD analysis did not reveal any extra phases other than ZnO, however, the HRTEM observation revealed that there was a $ZnAl_2O_4$ spinel phase formation both at grain boundaries and within the grains as seen in Fig. 5a. In other words, a small portion of Al was dissolved into ZnO as a donor and the rest reacted with ZnO to form $ZnAl_2O_4$, which is consistent with the report from Tsubota[15]. The periodical structures were also observed at grain boundary as seen in Fig. 5(b) and 5(c). A rotation angle at grain boundary in Fig. 5(b) and (c) is estimated about $16.5°$ and 15-$16°$, respectively. The $16.5°$ rotation angle corresponded to $\Sigma49$ grain boundary[16]. These periodical structures observed are due to the results of magnetic alignment between the grains perpendicular to c-axis. These unique structures correspond to a lower energy state that does not become the origin of double Shottky barrier in pure ZnO, hence, the electron scattering at the grain boundary would be less dominant. On the other hand, no particular crystal structure at grain boundaries in the randomly oriented specimen was found. The corresponding lattice mismatch, site defect, or dislocation hindered the electrical conductivity. As a result, we found that the morphology of grain boundary directly influenced the electrical conductivity.

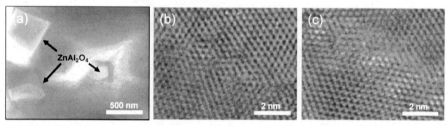

Figure 5. HRTEM images of the magnetically aligned specimen.

CONCLUSIONS

We successfully fabricated the highly c-axis oriented Al-ZnO ceramics by applying the rotational high magnetic field via gelcasting technique. The particles were aligned within a short time along the c-axis and then locked by the gelation, and the particle orientation was not degraded by the gelation and subsequent processing. The elongated grains suggested that the growth of the ab-plane enhances the degree of orientation during the sintering, realizing a highly oriented specimen. The c-axis specimen along the ab-plane showed a higher electrical conductivity than that along the c-axis while the Seebeck coefficient and thermal conductivity showed a small anisotropy. HRTEM observation revealed that the coincidence site lattices were often observed in the oriented specimen, suggesting that the unique microstructure was believed to be the main cause of these transport differences.

REFERENCES

[1]Y. Saito, H. Takao, T. Tani, T. Nonoyama, K. Takatori T. Homma, T. Nagaya and M. Nakamura, "Lead-free Piezoceramics," *Nature*, **432**, 84-87 (2004).
[2]T. Takenaka and K. Sakata, "Grain Orientation and Electrical Properties of Hot-Forged Bi₄Ti₃O₁₂ Ceramics," *J. Appl. Phys.*, **19**, 31-39 (1980).
[3]K. Hirao, M. Ohashi, M. E. Brito and S. Kanzaki, "Further Improvement in Mechanical Properties of Highly Anisotropic Silicon Nitride Ceramics," *J. Am. Ceram. Soc.*, **83**, 495-00 (2000).
[4]M. W. Seabaugh, I. H. Kerscht and G. L. Messing, "Texture Development and Microstructure Evolution in Liquid-Phase-Sintered α-Alumina Ceramics Prepared by Templated Grain Growth," *J. Am. Ceram. Soc.*, **83**, 3109-16 (2000).
[5]T. Tani, "Piezoelectric Properties of Bismuth Layer-Structured Ferroelectric Ceramics with a Preferred Orientation Processed by the Reactive Templated Grain Growth method," *J. Korean Phys. Soc.*, **32**, S1217-21 (1998).
[6]P. De. Rango, M. Lees, P. Lejay, A. Sulpice, R. Tournier, M. Ingold, P. Germi and M. Pernet, "Texturing of Magnetic Materials at High Temperature by Solidification in a Magnetic Field," *Nature*, **349**, 770-72 (1991).
[7]H. Sata, T. Kimura, S. Ogawa, M. Yamato and E. Ito, "Magnetic Orientation of Poly(ethylene-2,6-naphthalate) ," *Polymer*, **37**, 1879-82 (1996).

[8]S. Tanaka, A. Makiya, Z. Kato, N. Uchida, T. Kimura, K. Uematsu, "Fabrication of *c*-axis Oriented Polycrystalline ZnO by Using a Rotating Magnetic Field and Following Sintering," *J. Mater. Res.*, **21**, 703-07 (2006).

[9]H. Kaga, Y. Kinemuchi, S. Tanaka, A. Makiya, Z. Kato, K. Uematsu, K. Watari, "Preparation and Thermoelectric Property of Highly Oriented Al-Doped ZnO Ceramics by a High Magnetic Field," *Jpn. J. Appl. Phys.*, **45**, L1212-14 (2006).

[10]Y. Sakka and T. S. Suzuki, "Textured Development of Feeble Magnetic Ceramics by Colloidal Processing under High Magnetic Field," *J. Ceram. Soc. Jpn.*, **113**, 26-36 (2005).

[11]Y. Nakamura, T. Harada, H. Kuribara, A. Kishimoto, Motohara, H. Yanagida, "Non-linear, Current-Volatage Characteristics with Negative Resistance Observed at ZnO-ZnO Single Contacts," *J. Am. Ceram. Soc.*, **82**, 3069-74 (1999).

[12]T. S. Suzuki and Y. Sakka, "Control of Texture in ZnO by Slip Casting in a Strong Magnetic Field and Heating ," *Chem. Lett.*, 1204-07 (2002).

[13]A. C. Young, O. O. Omatete, M. A. Janney and P. A. Menchofer, "Gelcasting of Alumina," *J. Am. Ceram. Soc.*, **74**, 612-18 (1991).

[14]W. A. Dollase, "Correction of Intensities for Preferred Orientation in Powder Diffractometry: Application of the March Model," *J. Appl. Cryst.*, **19**, 267-72 (1986).

[15]T. Tsubota, M. Ohtaki, K. Eguchi, H. Arai, "Thermoelectric Properties of Al-Doped ZnO as a Promising Oxide Material for High-Temperature Thermoelectric Conversion," *J. Mater. Chem.*, **7**, 85-90 (1997).

[16]Y. Sato, T. Mizoguchi, F. Oba, M. Yodogawa, T. Yamamoto, and Y. Ikuhara, "Atomic and Electronic Structure of [0001]/([1230]) Σ7 Symmetric Tilt Grain Boundary in ZnO Bicrystal with Linear Current-Voltage Characteristic," *J. Mater. Sci.*, **40**, 3059-66 (2005).

POROUS ANATASE TITANIUM DIOXIDE FILMS PREPARED IN AQUEOUS SOLUTION

Yoshitake Masuda, Tatsuo Kimura, Kazumi Kato and Tatsuki Ohji
National Institute of Advanced Industrial Science and Technology (AIST)
2266-98 Anagahora, Shimoshidami, Moriyama-ku, Nagoya 463-8560, Japan

ABSTRACT

Porous anatase TiO_2 films were deposited on SnO_2: F substrates in an aqueous solution containing ammonium hexafluorotitanate ($[NH_4]_2TiF_6$) and boric acid (H_3BO_3) at 50 °C. The thickness was increased as deposition time. The films were constructed of multi-needle particles. Surface of the films showed large roughness due to nano/micro-asperity of the particles. The films showed high intensity of 0004 X-ray diffraction because the needles elongated along the c-axis. Surface morphology of the constituent TiO_2 particles contributed to high specific surface area of the films. The porous TiO_2 film has a great deal of potential in the development of biomolecule sensors.

INTRODUCTION

Titanium dioxide (TiO_2) thin films are of interest for various applications including microelectronics[1], optical cells[2], solar energy conversion[3], highly efficient catalysts[4], microorganism photolysis[5], antifogging and self-cleaning coatings[6], gratings[7], gate oxides in MOSFETs (Metal-Oxide-Semiconductor Field Effect Transistor)[8], etc. Accordingly, various attempts have been made to fabricate thin films and micropatterns of TiO_2, and in particular, the method based on an aqueous solution is important as an environment-friendly synthesis process, i.e., "green chemistry".

TiO_2 films have been prepared from aqueous solutions via various methods[9-11]. Deki et al.[9,10] reported the preparation of amorphous TiO_2 thin films containing 37 vol%[10] nonoriented anatase TiO_2 on glass substrates at 30°C from $(NH_4)_2TiF_6$ aqueous solution using liquid phase deposition which was first used for SiO_2[12]. The TiO_2 thin films were transparent because they were made of polycrystalline particles with diameters smaller than the wavelength of visible light. Recently, anatase TiO_2 thin films have been successfully fabricated at 50°C using liquid phase deposition (crystallization)[13]. The nucleation and growth process of anatase TiO_2 on several kinds of self-assembled monolayers (SAMs) in an aqueous solution has been evaluated using a quartz crystal microbalance[13]. Homogeneously nucleated TiO_2 particles and amino groups of SAM showed negative and positive zeta potential in the solution, respectively. The adhesion of TiO_2 particles to the amino group at the surface by attractive electrostatic interaction would cause rapid growth of TiO_2 thin films in the supersaturated solution at pH 2.8. On the other hand, TiO_2 was deposited on SAMs without the adhesion of TiO_2 particles regardless of the type of SAM in the solution at pH 1.5 whose supersaturation degree was low due to high concentration of H^+. The growth via attachment of particles is usually associated with high growth rate, but also high roughness and reduced crystallographic orientation. However, the method described above avoided this compromise, yielding a high growth rate as well as partial crystallographic orientation and smooth surfaces. Additionally, liquid phase patterning (LPP) of anatase TiO_2 was realized using SAMs[14]. It was proposed based on scientific knowledge obtained from investigations of interactions and chemical reactions between functional groups of SAMs and ions, clusters and homogeneously nucleated particles in solutions. Mechanisms and site-selectivities for LPP were also discussed in details. These studies showed high performance and high potential of solution chemistry for inorganic materials.

In this study we fabricated porous anatase TiO_2 film on SnO_2: F (FTO) substrates using LPD. Morphology of TiO_2 particles and particulate films was controlled to have high specific surface area. Film thickness and morphology were investigated as a function of deposition time. The anatase TiO_2 film on SnO_2: F (FTO) substrates developed in this study can be applied to dye-sensitized solar cells and biomolecule sensors.

EXPERIMENTAL

Ammonium hexafluorotitanate ($[NH_4]_2TiF_6$) (Morita Chemical Industries Co., Ltd., FW: 197.95, purity 96.0%) and boric acid (H_3BO_3) (Kishida Chemical Co., Ltd., FW: 61.83, purity 99.5%) were used as received. Silicon wafers or a glass substrate coated with F doped SnO_2 transparent conductive film (FTO, SnO_2: F, Asahi Glass Co., Ltd., 9.3-9.7 $\Omega/\square$, 26 × 50 × 1.1 mm) were used after surface cleaning. Both ends (26 × 14 mm) of the glass substrates were covered by scotch tape (CM-18, 3M) to prevent deposition. Morphology of TiO_2 films was observed by a field emission scanning electron microscope (FE-SEM; JSM-6335F, JEOL Ltd.) and a transmission electron microscope (TEM; JEM4010, 400 kV, point-to-point resolution 0.15 nm, JEOL Co., Ltd.). The crystal phase was evaluated by an X-ray diffractometer (XRD; RINT-2100V, Rigaku) with CuKα radiation (40 kV, 30 mA). The diffraction patterns were evaluated using ICSD (Inorganic Crystal Structure Database) data (FIZ Karlsruhe, Germany and NIST, USA) and FindIt.

Ammonium hexafluorotitanate (2.0096 g) and boric acid (1.86422 g) were separately dissolved in deionized water (100 mL) at 50°C. An appropriate amount of HCl was added to the boric acid solution to control pH. Boric acid solution was added to ammonium hexafluorotitanate solution. The concentrations of boric acid solution and ammonium hexafluorotitanate were 0.15 M and 0.05 M, respectively. Solutions (200 ml) with 0, 0.2 or 1.2 ml of HCl showed pH 3.8, 2.8 or 1.5, respectively. Supersaturation of solution can be changed by pH value. Silicon wafers or FTO substrates were immersed vertically in the middle of the solution[13]. The solution was kept at 50°C for 48 h. The solution became cloudy 10 min after the mixing of the solutions. The particles were homogeneously nucleated in the solution and made the solution white. They were then gradually precipitated and covered the bottom of the vessel over a period of several hours. The substrates were removed from the solution in 2, 5, 25 and 48 h.

RESULTS AND DISCUSSION
(1) Deposition of anatase TiO_2

Deposition of anatase TiO_2 proceeds by the following mechanisms[13] (Fig. 1):

$$TiF_6^{2-} + 2H_2O \rightleftharpoons TiO_2 + 4H^+ + 6F^- \quad ...(a)$$
$$BO_3^{3-} + 4F^- + 6H^+ \longrightarrow BF_4^- + 3H_2O \quad ...(b)$$

Equation (a) is described in detail by the following two equations:

$$TiF_6^{2-} \xrightarrow{nOH^-} TiF_{6-n}(OH)_n^{2-} + nF^- \xrightarrow{(6-n)OH^-} Ti(OH)_6^{2-} + 6F^- \quad ...(c)$$
$$Ti(OH)_6^{2-} \longrightarrow TiO_2 + 2H_2O + 2OH^- \quad ...(d)$$

$$TiF_6^{2-} + 2H_2O \rightleftharpoons TiO_2 + 4H^+ + 6F^-$$

$$BO_3^{3-} + 4F^- + 6H^+ \longrightarrow BF_4^- + 3H_2O$$

$$TiF_6^{2-} \xrightarrow{\ nOH^-\ } TiF_{6-n}(OH)_n^{2-} + nF^- \xrightarrow{\ (6-n)OH^-\ } Ti(OH)_6^{2-} + 6F^-$$

$$Ti(OH)_6^{2-} \xrightarrow{\ H_2O\ } TiO_2$$

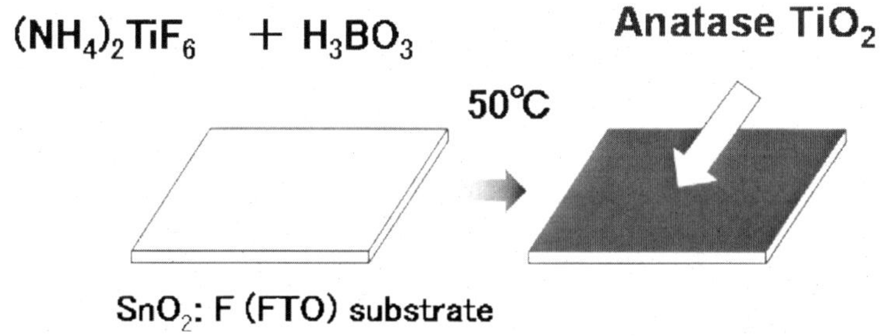

$(NH_4)_2TiF_6 \quad + \quad H_3BO_3$ 　　　　　　**Anatase TiO$_2$**

50°C

SnO$_2$: F (FTO) substrate

Figure 1. Conceptual process for fabricating anatase TiO$_2$ thin film from an aqueous solution.

Fluorinated titanium complex ions gradually change into titanium hydroxide complex ions in the aqueous solution as shown in Eq. (c). Increase of F$^-$ concentration displaces the Eq. (a) and (c) to the left, however, produced F$^-$ can be scavenged by H$_3$BO$_3$ (BO$_3^{3-}$) as shown in Eq. (b) to displace the Eq. (a) and (c) to the right. Anatase TiO$_2$ was formed from titanium hydroxide complex ions (Ti(OH)$_6^{2-}$) in Eq. (d), and thus the supersaturation degree and the deposition rate of TiO$_2$ depend on the concentration of titanium hydroxide complex ions. The high concentration of H$^+$ displaces the equilibrium to the left in Eq. (a), and the low concentration of OH$^-$, which is replaced with F$^-$ ions, suppresses ligand exchange in Eq. (c) and decreases the concentration of titanium hydroxide complex ions at low pH such as pH 1.5. The solution actually remained clear at pH 1.5, showing its low degree of supersaturation. On the other hand, the solution at high pH such as pH 2.8 or 3.8 became turbid because of homogeneously-nucleated anatase TiO$_2$ particles caused by a high degree of supersaturation. Anatase TiO$_2$ thin film was formed by heterogeneous nucleation in the solution at pH 1.5, while the film was formed by heterogeneous nucleation and deposition of homogeneously nucleated particles at pH 2.8 or 3.8.

(2) Crystal phase of as-deposited films[15]

After having been immersed in the solution at pH 3.8, the substrates with films were dried in air. The films were colored to slight white. This indicated that the films were deposited with no colored by-product. The films showed the same whiteness over the whole area, which supported the high uniformity of film thickness and chemical composition. Whiteness gradually increased as a function of deposition time due to increase of film thickness. The films did not peel off during ultrasonic oscillation treatment in acetone for 30 min as they showed high adhesion strength.

Strong X-ray diffractions were observed for films deposited on FTO substrates and assigned to SnO_2 of FTO films. Glass substrates with no FTO coating were immersed in the solution for comparison. Weak X-ray diffraction peaks were observed at $2\theta = 25.3$, 37.7, 48.0, 53.9, 55.1 and 62.7° for films deposited on glass substrates and assigned to 101, 004, 200, 105, 211 and 204 diffraction peaks of anatase TiO_2 (ICSD No. 9852), respectively. (Fig. 2). A broad diffraction peak from the glass substrate was also observed at about $2\theta = 25°$. The 004 diffraction peak of anatase TiO_2 was not distinguished clearly for film on FTO substrates because both of the weak 004 diffraction peak of TiO_2 and the strong diffraction peak of FTO were observed at the same angle. Crystallite size perpendicular to (004) planes was estimated from the full-width half-maximum of the 004 peak to be 17 nm.

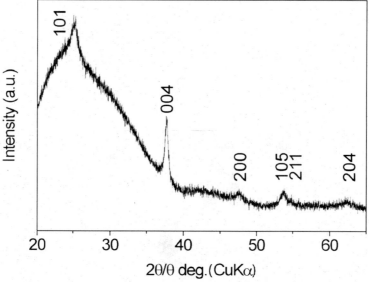

Figure 2. XRD diffraction pattern of anatase TiO_2 film on glass substrate.

(3) Morphology and film thickness change of TiO₂ films[15]

Thickness of film increased to 260 nm by the immersion for 2 h at pH 3.8. The TiO_2 films were constructed of particles and had a relief structure on the surface (Fig. 3a). The diameter of the particles was estimated to be 100–600 nm and cracks were observed at the boundaries of particles, which would have been generated during the drying process due to shrinkage of the films. The particles had many projections on the surface, and so probably had a multi-needle shape (Fig. 3b). Thus, the films had a large relief structure due to the assembly of particles and a small relief structure due to needles on the particle surface.

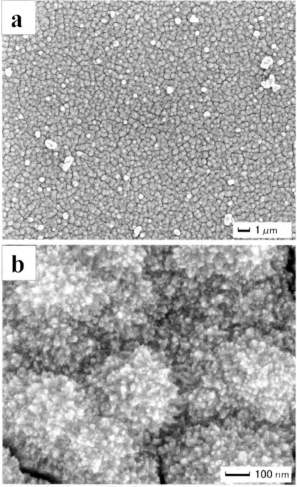

Figure 3. SEM micrographs of anatase TiO_2 films deposited on SnO_2: F substrates for 2 h. (a) Surface of anatase TiO_2 films. (b) Magnified area of (a) showing morphology of anatase TiO_2 particles.

The films grew to 360 nm, 600 nm and 760 nm in thickness for 5, 25 and 48 h, respectively. The surface of the films gradually became smoother as a function of deposition time (Fig. 4a, Fig. 5a, Fig. 6a).

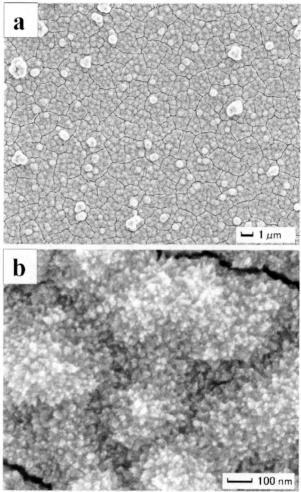

Figure 4. SEM micrographs of anatase TiO_2 films deposited on SnO_2: F substrates for 5 h. (a) Surface of anatase TiO_2 films. (b) Magnified area of (a) showing morphology of anatase TiO_2 particles.

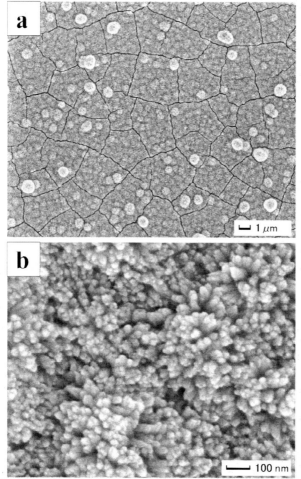

Figure 5. SEM micrographs of anatase TiO_2 films deposited on SnO_2: F substrates for 25 h. (a) Surface of anatase TiO_2 films. (b) Magnified area of (a) showing morphology of anatase TiO_2 particles.

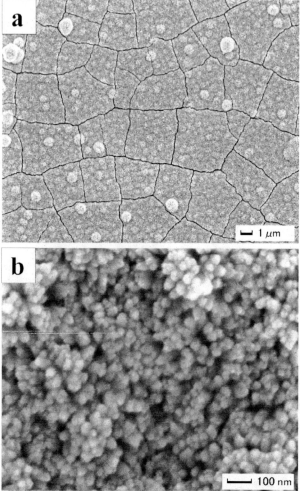

Figure 6. SEM micrographs of anatase TiO₂ films deposited on SnO₂: F substrates for 48 h. (a) Surface of anatase TiO₂ films. (b) Magnified area of (a) showing morphology of anatase TiO₂ particles.

Black and white contrast caused from surface relief in the micrographs decreased gradually. On the other hand, the size of cracks increased with deposition time. Thick films would suppress the generation of small cracks to form long cracks because of stress accumulation due to the high mechanical strength of the films. Boundaries of particles were modified and became more difficult to determine as deposition time increased because of film growth. Particle size was roughly estimated to be 300–600, 450–600 and 550–670 nm for 5, 25, and 48 h, respectively. On the other hand, small needles on the surface of films changed and became visibly clearer as deposition time increased (Fig. 4b, Fig. 5b, Fig. 6b). Small needles of anatase crystals increased in size. Rapid growth occurred at the initial stage and the growth rate gradually decreased with deposition time.

Figure 7 shows TEM micrograph and electron diffraction pattern for the cross section profile of TiO$_2$ thin film deposited on silicon substrate. pH of the solution was adjusted to pH 2.8 by addition of HCl. Many small crystals of anatase TiO$_2$ were observed throughout the thin film. The electron diffraction pattern also showed weak c-axis orientation of anatase TiO$_2$ crystals.

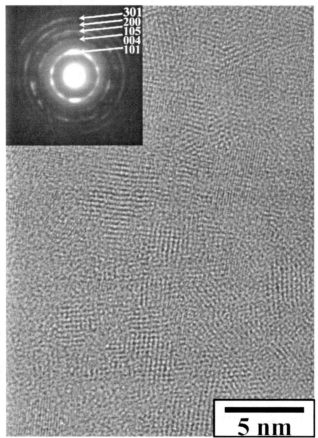

Figure 7. TEM micrograph and electron diffraction pattern for cross section profile of TiO$_2$ thin film deposited at pH 2.8.

The particles of anatase TiO_2 were generated homogeneously as observed for white turbidity of the solution. The particles grew further and increased in size, gradually precipitating on the bottom of the vessel. They also adhered to the substrates which stood in the middle of the solution. The adhesion of the particles would cause rapid growth of the films. Heterogeneous nucleation would also occur on the surface of the substrates. The thickness of the particulate films increased due to rapid crystal growth at the initial stage, in which the ion concentration was sufficiently high for heterogeneous film growth. Ion concentration would decrease as a function of deposition time because of the formation and precipitation of TiO_2 particles and the film growth. Film growth would be driven mainly by heterogeneous crystal growth after 2 h. The particulate films grew to form a smooth surface and particle size increased when observed from perpendicular to the surface, blurring the boundaries between the particles. Consequently, rough particulate films grew, increased in thickness and had a smooth surface with nano-scale relief structure.

CONCLUSION

Crystalline anatase TiO_2 was deposited from the solution containing ammonium hexafluorotitanate and boric acid at 50°C. Amorphous TiO_2 was formed by liquid phase deposition at 30°C, however, TiO_2 was crystallized by the increase of solution temperature. Liquid phase crystallization of TiO_2 was thus successfully achieved in this system. Anatase TiO_2 films were prepared on FTO substrates at pH 3.8. The particles were homogeneously nucleated and adhered to the substrates at the initial stage to form particulate films. The films were then grown with deposition period. Small needles of anatase TiO_2 crystals on the surface of the films increased in size and so became visibly clearer as deposition time increased. Thus, the films had a large relief structure due to the assembly of particles and a nano-scale relief structure due to needles on the particle surface. The anatase TiO_2 films would be expected to apply as porous electrodes for next-generation biomolecule sensors and dye-sensitized solar cells.

REFERENCES

[1] G. P. Burns, "Titanium-dioxide Dielectric Films Formed by Rapid Thermal-oxidation," J. Appl. Phys. **65**, 2095-2097 (1989).

[2] B. E. Yoldas and T. W. O'Keeffe, "Anti-reflective Coatings Applied from Metal-organic Derived Liquid Precursors," Appl. Opt. **18** (18), 3133-3138 (1979).

[3] M. A. Butler and D. S. Ginley, "Principle of Photoelectrochemical, Solar-energy Conversion," J. Mat. Sci. **15** (1), 1-19 (1980).

[4] T. Carlson and G. L. Giffin, "Photooxidation of Methanol using V_2O_5/TiO_2 and MoO3/TiO2 Surface Oxide Monolayer Catalysts," J. Phys. Chem. **90** (22), 5896-5900 (1986).

[5] T. Matsunaga, R. Tomoda, T. Nakajima et al., "Continuous-sterilization System that uses Photosemiconductor Powders," Appl. Environ. Microbiol. **54** (6), 1330-1333 (1988).

[6] R. Wang, K. Hashimoto, and A. Fujishima, "Light-induced amphiphilic surfaces," Nature **388** (6641), 431-432 (1997).

[7] S. I. Borenstain, U. Arad, I. Lyubina et al., "Optimized random/ordered grating for an n-type quantum well infrared photodetector," Thin Solid Films **75** (17), 2659-2661 (1999).

[8] P. S. Peercy, "The drive to miniaturization," Nature **406** (6799), 1023-1026 (2000); D. J. Wang, Y. Masuda, W. S. Seo et al., "Metal-oxide-semiconductor (MOS) devices

composed of biomimetically synthesized TiO$_2$ dielectric thin films," Key Eng. Mater. **214** (2), 163-168 (2002).

[9] S. Deki, Y. Aoi, O. Hiroi et al., "Titanium(IV) oxide thin films prepared from aqueous solution," Chem. Letters **6**, 433-434 (1996); S. Deki, Y. Aoi, H. Yanagimoto et al., "Preparation and characterization of Au-dispersed TiO2 thin films by a liquid-phase deposition method," J. Mater. Chem. **6** (12), 1879-1882 (1996); S. Deki, Y. Aoi, Y. Asaoka et al., "Monitoring the growth of titanium oxide thin films by the liquid-phase deposition method with a quartz crystal microbalance," J. Mater. Chem. **7** (5), 733-736 (1997).

[10] H. Kishimoto, K. Takahama, N. Hashimoto et al., "Photocatalytic activity of titanium oxide prepared by liquid phase deposition (LPD)," J. Mater. Chem. **8** (9), 2019-2024 (1998).

[11] Q. Chen, Y. Qian, Z. Chen et al., "Low-temperature deposition of ultrafine rutile TiO2 thin films by the hydrothermal method," Phys. Stat. Sol. (a) **156** (2), 381-385 (1996); D. Huang, Z-D. Xiao, J-H. Gu et al., "TiO$_2$ thin films formation on industrial glass through self-assembly processing," Thin Solid Films **305** (1-2), 110-115 (1997); K. Shimizu, H. Imai, H. Hirashima et al., "Low-temperature synthesis of anatase thin films on glass and organic substrates by direct deposition from aqueous solutions," Thin Solid Films **351** (1-2), 220-224 (1999); X P Wang, Y Yu, X F Hu et al., "Hydrophilicity of TiO$_2$ films prepared by liquid phase deposition," Thin Solid Films **371** (1-2), 148-152 (2000); M. K. Lee and B. H. Lei, "Characterization of titanium oxide films prepared by liquid phase deposition using hexafluorotitanic acid," Jpn. J. Appl. Phys. **39** (2A), L101-103 (2000); U. Selvaraj, A. V. Prasadarao, S. Komarneni et al., "Sol-gel Fabrication of Epitaxial and Oriented TiO$_2$ Thin-films," J. Am. Ceram. Soc. **75** (5), 1167-1170 (1992).

[12] H. Nagayama, H. Honda, and H. Kawahara, "A New Process for Silica Coating," J. Electrochem. Soc. **135** (8), 2013-2016 (1988).

[13] Y. Masuda, T. Sugiyama, W. S. Seo et al., "Deposition Mechanism of Anatase TiO$_2$ on Self-Assembled Monolayers from an Aqueous Solution," Chem. Mater. **15** (12), 2469-2476 (2003).

[14] Y. Masuda, S. Ieda, and K. Koumoto, "Site-Selective Deposition of Anatase TiO$_2$ in an Aqueous Solution Using a Seed Layer," Langmuir **19** (10), 4415-4419 (2003); Y. Masuda, T. Sugiyama, and K. Koumoto, "Micropattening of anatase TiO$_2$ thin films from an aqueous solution by site-selective immersion method," J. Mater. Chem. **12** (9), 2643-2647 (2002); Y. Masuda, N. Saito, R. Hoffmann et al., "Nano/micro-patterning of anatase TiO$_2$ thin film from an aqueous solution by site-selective elimination method," Sci. Tech. Adv. Mater. **4**, 461-467 (2003).

[15] Y Masuda and K. Kato, Thin Solid Films, submitted.

PREPARATION OF MICRO/MESOPOROUS Si-C-O CERAMIC DERIVED FROM PRECERAMIC ROUTE

Manabu FUKUSHIMA, You ZHOU, Yu-ichi YOSHIZAWA, Masayuki NAKATA and   Kiyoshi HIRAO

National Institute of Advanced Industrial Science and Technology (AIST), 2266-98 Shimo-Shidami, Moriyama-ku, Nagoya, Aichi, 463-8560, Japan

ABSTRACT
    Micro/mesoporous silicon oxycarbide films were prepared by the pyrolysis of preceramic polymer with and without filler. In this study, SiC particle filler to prevent shrinkage was used and its effect on surface area was investigated. The pyrolyzed film without filler showed much lower surface area around 3-5m^2/g, while the film with filler had higher surface area around 150-300m^2/g. This is due to the inhibition of shrinkage of film during pyrolysis, namely the closure of micro/mesopores by the addition of filler.

INTRODUCTION

    Porous ceramic filters have been widely utilized for hydrogen separation-production system by the steam modification of natural gas, filtration to reuse polishing slurry and water purification plant, because of their high thermal and chemical stability as well as excellent mechanical property [1-8]. In addition, the pore size distribution of porous ceramic materials is narrower than that of organic materials, which is an important factor to dominate separation ability. Macroporous ceramic is generally prepared by partial sintering at lower sintering temperature and with slight additive [1-3], while micro/meso porous ceramic is derived from precursor route as sol-gel or inorganic polymer [4-8]. A microporous (Φ < 2 nm) top membrane layer is coated on a mesoporous intermediate layer (2 nm < Φ < 50 nm) with a macroporous (50 nm < Φ) membrane support, providing an asymmetric structure with a pore size gradient. For the membrane system, alumina is used as a macroporous support, while silica or γ-alumina is used as micro/meso porous layer.

    Compared to those metastable oxide materials, silicon oxycarbide has better thermal shock resistance and thermal or chemical stability [9-10]. This strongly suggests that silicon oxycarbide is useful as a membrane or intermediate layer in separation system at elevated temperatures or corrosive environment. Silicon oxycarbide is generally prepared by the pyrolysis of polysiloxane through organically modified silane precursor. To increase porosity of oxycarbide, the foam by polyurethane with activated carbon and aging in ammonium hydroxide have been reported [11-14]. In addition, the pyrolysis temperature of preceramic polymer is an important parameter to provide pore by gas evolution and control surface area. Surface area is found to connect with permeation in filter application.

    Gas evolution of siloxane based polymer starts around 600°C. However, organic groups in pyrolyzed siloxane around 600°C remain, which leads to the lower thermal stability. Higher temperature pyrolysis causes the closure of micro/meso pores by dimensional changes due to shrinkage, resulting in the dramatic decrease of surface area. Namely, high temperature heating without shrinkage is important issue to provide high surface area and thermal stability. In this study, inorganic passive filler is used in order to inhibit the shrinkage at high temperature heating, and the relationship between pyrolysis temperature and surface area are investigated for the materials with and without filler.

EXPERIMENTAL PROCEDURE

Methyltriethoxysilane (MTES; Shin-Etsu Chemical Co., Ltd., Tokyo, Japan) was used as precursor polymer. High purity nano-sized β-SiC powder (Sumitomo Osaka Cement Co. Ltd., Tokyo, Japan) with an average particle size of 30 nm and a specific surface area of 40-50 m^2/g was used as filler. This powder was prepared by plasma CVD method, and composed of particles with spherical shape. The main impurities in the nano-sized SiC powder were 1.8-2.4 mass% carbon and 0.3-0.6 mass% oxygen. As inorganic component, chemically modified titanium tetraisopropoxide with acetylacetone (TIP; Wako Pure Chemical Industries, Ltd., Tokyo, Japan) was used. These mixtures with the weight ratios of 0.05 = SiC/(MTES+SiC), TIP/MTES= 0.05/1, H_2O/MTES=2/1 and HCl/MTES=0.005/1 were blended in ethanol for 1 hr using a planetary mill with a SiC pot and SiC balls. After cast and drying under ambient atmosphere, polymethylsilsesquioxane (PMSQ) self-supported gel films with and without filler were obtained. Hereafter, these compositions are referred to SiC/Ti-PMSQ (with filler) and Ti-PMSQ (without filler). The green samples preheated at 200°C were placed in a quartz dish and further heated at 400, 600 and 800°C for 2 hr under Ar gas flow. The heating rate from room temperature to setting temperatures was 3°C/min. The shrinkage of pyrolyzed films was measured by 10 points average values of micrometer. The specific surface area of pyrolyzed films was characterized by nitrogen adsorption measurement (Yuasa Ionics Inc., Autosorb, Osaka, Japan). BET surface area was determined from a BET (Brunauer, Emmet and Teller) analysis in the P/P_0 range of 0.05-0.30 using a molecular cross sectional area for N_2 of 0.163 nm^2 and multi points.

RESULTS AND DISCUSSION

Table 1 shows linear shrinkage as a function of pyrolysis temperature. The effect of pyrolysis temperature was clearly observed; shrinkage increased with increase of pyrolysis temperature. In addition to pyrolysis temperature, the effect of filler on shrinkage was also monitored. The shrinkage of pyrolyzed SiC/Ti-PMSQ films was always 1-2% lower values than that of SiC/Ti-PMSQ. The addition of filler was found to reduce shrinkage during pyrolysis.

Table 2 illustrates typical result of ceramic yield of Ti-PMSQ film as a function of pyrolysis temperature. Weight loss was in the range of 2-11%. The weight loss was found to be larger around 600-800°C, which may be connected with gas evolution during pyrolysis of PMSQ. Large amount of gas evolution is considered to result in large shrinkage, because the pore formed by gas evolution can easily close to reduce its surface energy. Actually, larger shrinkage occurred at 800°C.

Figure 1 shows the FT-IR spectra of pyrolyzed Ti-PMSQ films at various temperatures. The adsorption bands around 910-930 (Si-O-Ti) and 1275 (Si-CH$_3$) cm^{-1} decreased with increase of pyrolysis temperature. When pyrolyzed at 800°C, these bands disappeared. This means that the dispersed titania moved out of silsesquioxane network and terminal methyl groups decomposed. The cleavage of methyl groups can be accompanied by gas evolution such as hydrogen, methane and ethane. The evolution of bulky hydrocarbon gas should be related with the formation of pore and shrinkage.

Figure 2 shows the specific surface area of the pyrolyzed Ti-PMSQ and SiC/Ti-PMSQ at 400-800°C and pure SiC filler. The surface area of SiC/Ti-PMSQ film showed the maximum values of 309m^2/g, when pyrolyzed at 600°C. The surface area of the pyrolyzed SiC/Ti-PMSQ at

800°C was 147 m²/g and lower than that at 600°C. In contrast, the surface area of Ti-PMSQ and pure filler showed constant values around 3-5 and 50m²/g, respectively. Pure filler can not sinter in this temperature range. Namely, the filler itself does not have high surface area but affects the surface area of Ti-PMSQ matrix in SiC/Ti-PMSQ film.

Table 1 Linear shrinkage as a function of pyrolysis temperature

Pyrolysis	Shrinkage (%)	
temperature (°C)	Ti-PMSQ	SiC/Ti-PMSQ
400	4	3
600	5	4
800	11	9

Table 2 Ceramic yield as a function of pyrolysis temperature

	Pyrolysis temperature (°C)		
	400	600	800
Ceramic yield (%)	98	94	89

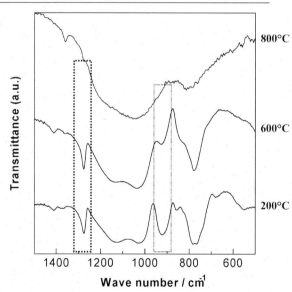

Figure 1 FT-IR spectra of pyrolyzed Ti-PMSQ films at various temperatures.

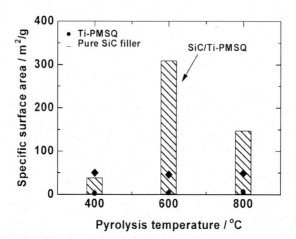

Figure 2 Specific surface area of pyrolyzed SiC/Ti-PMSQ and Ti-PMSQ films at various temperatures and pure filler.

We compared the surface area observed in our study with those of Schmidt and Walter [11, 15], where they studied the surface area of pyrolyzed polysiloxane with activated carbon or its foam prepared by impregnation of polyurethane and pyrolyzed silicon oxycarbide with various silica contents, respectively. They found that surface area temporarily increased and decreased with increasing pyrolysis temperature, which was consistent with our results. In the case of report by Schmidt, the addition of activated carbon or polyurethane led to a larger surface area; while pure bulk siloxane had the surface area of 401 and 66m²/g for the specimens pyrolyzed at 600 and 800°C, the addition of activated carbon and polyurethane resulted in higher surface area of 426 and 138m²/g for specimens at 600 and 800°C, respectively [11]. In report by Walter, the surface area of oxycarbide increased with silica content; from 43 to 219m²/g when heated at 800°C [15]. Though macropores in our study could not be observed by mercury porosimetric analysis and additive with high surface area did not be used, pyrolysis temperature and surface area are found to have similar relation with above previous reports.

Nitrogen gas adsorption and desorption can show the properties of pore and surface area. The isotherms are illustrated in Fig.3. For the pyrolyzed SiC/Ti-PMSQ at 600°C, the increasing at low relative pressure in SiC/Ti-PMSQ, namely type I isotherm was monitored. The type I isotherm indicates the presence of micropore. In contrast, the other films did not show any adsorption.

The SiC/Ti-PMSQ film showed larger adsorption and higher surface area than that of Ti-PMSQ film. This means the effect of filler on nitrogen adsorption site, namely pore. Pore and surface area are connected with gas evolution during pyrolysis, because the evolved gas should provide pore and new surface. However, shrinkage can also occur to reduce surface energy. Its shrinkage means the closure of formed pore by gas evolution. Thus, when shrinkage occurs, surface area is reduced. As mentioned above, the addition of filler was a key factor to reduce shrinkage during pyrolysis, which could partially prevent the closure of pore formed by gas evolution and show high surface area.

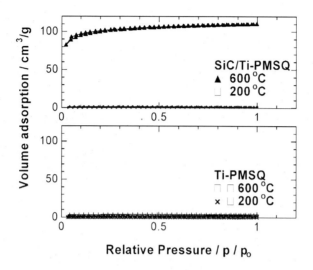

Figure 3 N$_2$ adsorption and desorption isotherms

CONCLUSION

Micro/mesoporous silicon oxycarbide films were prepared by the pyrolysis of preceramic polymer with and without filler. The pore properties such as pore size and surface area were investigated. The filler and the pyrolysis temperature significantly affected the surface area and pore size. The pyrolyzed film with filler showed higher surface area around 300 m^2/g because the shrinkage of film during pyrolysis, namely the closure of micro/mesopores was prevented by the addition of filler.

REFERENCES

[1] J. H. She, J. F. Yang, N. Kondo, T. Ohji and S. Kanzaki, "High-Strength Porous Silicon Carbide Ceramics by an Oxidation-Bonding Technique", J.Am.Ceram.Soc., 85, 11, 2852-2854 (2002).

[2] J.H.She, T. Ohji and S.Kanzaki, "Oxidation bonding of porous silicon carbide ceramics with synergistic performance", J.Eur.Ceram.Soc., 24, 331-334 (2003).

[3] M. Fukushima, Y. Zhou, Y. Iwamoto, S. Yamazaki, T. Nagano, H. Miyazaki, Y. Yoshizawa and K. Hirao, "Microstructural characterization of porous silicon carbide membrane support with and without alumina additive", J. Am. Ceram. Soc., 89, 1523-1529 (2006).

[4] H. Suda, H. Yamauchi, Y. Uchimaru, I. Fujiwara and K. Haraya, "Structural Evolution during Conversion of Polycarbosilane Precursor into Silicon Carbide-Based Microporous Membranes", J. Ceram. Soc. Japan, 114, 539-544 (2006).

[5] T. Nagano, K. Ssato, T. Saitoh and Y. Iwamoto, "Gas Permeation Properties of Amorphous

SiC Membranes Synthesized from Polycarbosilane without Oxygen-Curing Process", J. Ceram. Soc. Japan, 114, 533-538 (2006).

[6]M. Fukushima, Y. Zhou, Y. Yoshizawa and K. Hirao, "Preparation of Mesoporous Silicon Carbide from Nano-Sized SiC Particle and Polycarbosilane", J.Ceram.Soc.Jpn., 114, 571-574 (2006).

[7]Z. Li, K. Kusakabe and S. Morooka, "Preparation of thermostable amorphous Si-C-O membrane and its application to gas separation at elevated temperature", J. Membr. Sci., 118, 159-168 (1996).

[8]K. Kusakabe, Z. Li, H. Maeda and S. Morooka, "Preparation of supported composite membrane by pyrolysis of polycarbosilane for gas separation at high temperature", J. Membr. Sci., 103, 175-180 (1995).

[9]G. D. Soraru, S. Modena, E. Guadagnino, P. Colombo, J. Egan and C. Pantano, "Chemical Durability of Silicon Oxycarbide Glasses", J.Am.Ceram.Soc., 85, 6, 1529-1536 (2002).

[10]P. Colombo, J. R. Hellmann and D. L. Shelleman, "Thermal Shock Behavior of Silicon Oxycarbide Foams", J.Am.Ceram.Soc., 85, 9, 2306-2312 (2002).

[11]H. Schmidt, D. Koch, G. Grathwohl and P. Colombo, "Micro-/Macroporous Ceramics from Preceramic Precursors", J.Am.Ceram.Soc., 84, 10, 2252-2255 (2001).

[12]P. Colombo, E. Bernardo and L. Biasetto, "Novel Microcellular Ceramics from a Silicone Resin", J.Am.Ceram.Soc., 87, 1, 152-154 (2004).

[13]P. Colombo and M. Modesti, "Silicon Oxycarbide Ceramic Foams from a Preceramic Polymer", J.Am.Ceram.Soc., 82, 3, 573-578 (1999).

[14]A. K. Singh and C. G. Pantano, "Porous silicon oxycarbide glasses", J.Am.Ceram.Soc., 79, 10, 2696-2704 (1996).

[15]S. Waltera, G. D. Soraru, H. Brequel and S. Enzo, "Microstructural and mechanical characterization of sol gel-derived Si–O–C glasses", J.Eur.Ceram.Soc., 22, 2389-2400 (2002).

POROUS ALUMINA CERAMICS BY NOVEL GELATE-FREEZING METHOD

Masayuki Nakata, Manabu Fukushima, and Yu-ichi Yoshizawa

National Institute of Advanced Industrial Science and Technology (AIST)
2266 Anagahora, Shimo-Shidami
moriyama-ku, Nagoya, Aichi 463-8560
Japan

ABSTRACT

Porous alumina was fabricated by sintering green bodies formed by a novel gelate-freezing method using water dissolved polymer based slurries. These slurries were gelated by polymerization initiator, and frozen in a freezer at -20°C. During the drying process, moisture in the slurry was removed without vacuum deaeration. The fabricated porous alumina possessed continuous open pores, and a wide range of pore size and porosity. The pore volume fraction and the pore size were controllable by variation of alumina content in the slurry and particle size of alumina powder. The pore wall could be sintered to high densities, inside which only a few micro voids were observed.

INTRODUCTION

Porous ceramics are important materials for a variety of industrial applications such as liquid or gas filters [1], light materials, refractory tiles, catalyst carrier [2], sensors [3] and artificial bones. The pore sizes of porous materials are divided into 3 categories, macropore, mesopore and micropore (macropore>50nm>mesopore>2nm>micropore). Those materials are expected to have both higher fluid permeability and higher mechanical strength than porous materials with micro pores.

There are a variety of methods of fabricating ceramic forms with open pores, for examples: (1) Adjusting the density of ceramic perform and/or varying the sintering temperature or using reactive sintering, (2) Adding pore-forming agent such as polymer or carbon particles into the ceramic green body and then burning out the pore-forming agent by calcination or sintering, (3) Sintering a component consisting of dissoluble and indissoluble phases and then obtaining open pores by dissolving and removing the dissoluble phase, (4) impregnating ceramic slurry into a porous resin template and then removing the resin and sintering the preform. However, all these methods have some drawbacks such as the difficulty of controlling pore size distribution and the drastic drop in mechanical strength with increasing porosity. Although porous ceramics with relatively homogeneous distribution of pores and high porosities (>80%) can be prepared by using the method (4), cracks often occur in the frame-work structure due to thermal expansion or generation of thermo-decomposition gases, which result in extremely low mechanical strength.

In this paper, porous ceramics with structural function were fabricated from water solved polymer based ceramic slurry by freeze process. Alumina was used as a model material and the characteristics of the prepared porous ceramics bodies were evaluated in this study.

EXPERIMENTAL PROCEDURE

1. Materials

Fine-grained alumina powder (A160SG-4, Showa Denko Co., Ltd.), with a median size (d50) of 0.4 mm, specific surface area of 6.0 m^2/g, and density of 3.99 g/cm^3 was used. For a water soluble polymer solution, we used polyethylene-imine (50w% in H_2O, CAS 9002-98-6, Sigma). And Glycerol diglycidyl ether was used for gelling agent (CAS Number27043-36-3,

Aldrich). The most commonly used gel is acrylamide (AM) for producing ceramic green bodies by gelcasting. However, industry has been reluctant to use AM because of the neurotoxin.

Thus, we chose a low-toxicity gelcasting system by using a polyethylene-imine (PEI) based polymer. To produce stable alumina suspensions in water soluble polymer solution, we used a polycarboxylic acid ammonium solution (A6114, Toagosei Co., Ltd.) as a dispersant.

2. Fabrication Procedures

A flow chart of gelate-freezing method is shown in Fig.1. Suspensions were prepared by stirrer equipment (KEYENCE Co. Ltd., HM-500). Alumina powder, polymer solution and dispersant were mixed for 30 s. Slurries with initial solid contents of 10 and 20 vol% were prepared. The polymer concentration was always 5wt% to water. The dispersant concentration was always 1wt% to suspensions. The gelling agent was added to the mixed suspensions at a ratio of 0.5g agent to 10g of PEI and mixed for 30 sec by the stirring equipment. The slurries were poured into polyvinyl chloride molds at room temperature to produce a green body 23 mm in diameter.

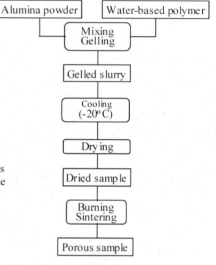

Fig 1. Flow chart of gelate-freeze method.

After casting, the suspension was kept in an oven at a temperature of 50 °C for 1 h for completion of the gelling. The resulted gelled slurry in the mold was frozen by isotropic cooling at -20 °C for 6 h in a freezer. After freezing process, frozen green bodies were dried in coordinate atmosphere. Removal of water in the sample was monitored by weighing the sample. The drying process typically finished in 24 h. After 90% of the water was removed, the sample was kept at 600 °C for 2 h in an electric tubular furnace to remove polymer from the dried green body.

The resulted sample was sintered at 1600 °C for 2 h.

CHARACTERIZATION

The porosity of sintered specimens was determined using the Archimedes method. The microstructure of the specimen was observed with scanning electron microscopy.

RESULTS

The slurry concentration, density and porosity of the obtained porous alumina sintered bodies are listed in Table 1. By decreasing the content of alumina, the porosity increased from 55% to 72%.

SEM microscopic photograph of porous alumina sintered with 55% porosity and 72% porosity was shown in Fig. 2 and Fig. 3. In Fig. 2 and Fig. 3, large pores with high aspect ratio (major axis: about 200µm, minor axis: about 30µm) were observed. The pores structure was surrounded by dense walls and had nearly elliptical cross-sections. The pore volume fraction and the pore size were controlled by the slurry concentration. Those porous were different from structures large flat ellipsoidal cross-sections[4-6] in conventional aqueous freeze casting.

Table 1. Porosity of porous alumina sintered bodies fabricated

No.	Slurry Concentration vol%)	Density (%)	Porosity (%)
1	20	1.70	57.3
2	20	1.78	55.3
3	10	1.13	71.7
4	10	1.17	69.7

Comparing Fig.2 and Fig. 3, it was clarified that shape of pore was sharpen and pore distribution was spread by decreasing slurry concentration. It is certified that porosity was decreased by decreasing slurry concentration.

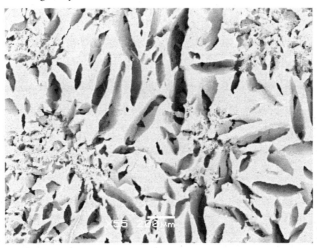

Fig.2 SEM microscopic photograph of porous alumina sintered with 55% porosity.

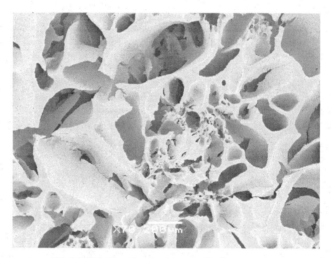

Fig.3 SEM microscopic photograph of porous
alumina sintered with 72% porosity.

DISCUSSION

In the present gelate-freezing method, ice crystals grew in the polymer during the freezing process. It was established that, micro-sized ice crystals grow inside when water solved polymers were frozen. These ice crystals grown in situ could easily to remove by freeze-drying process. Therefore, pores in the samples were made from such ice crystals. After the successive sintering process, porous sintered bodies were prepared. Ice growth is important for both pore size and shape in sintered bodies. The porosity and the pore shape in the sintered bodies were dependent of the water content in the slurry. Alumina particles attached to each other in the polymer solution, therefore the gelate-freezing method was able to use low concentration of slurry.

The present method is considered to be a freeze-dry method[4-6]. For comparison, the characteristic features of porous materials fabricated by the gelate-freezing method, the pore size fabricated by the present method is controllable and is smaller than that fabricated by the freeze-dry method. In the gel-casting method, it is known that porous ceramics with dense structures are fabricated by using fine ceramics powders. Desired features of porous sintered alumina, such as high porosity, continuous open pores, dense structure, and a wide range of pore sizes were achieved with our method.

CONCLUSION

In this study, porous alumina sintered bodies were fabricated by gelled freeze process. Using of gelled freeze process of water based polymer ceramics slurry, the porous alumina sintered bodies with dense framework structure and large pore size was fabricated successfully. The pore size and porosity were controllable by the slurry concentration. The fabricated porous ceramics had highly continuous pore and dense alumina part.

REFERENCES

[1] J. M. Almanza R, A. H. Castillejos E, F. A. Acosta G and A. Flores V, "Microstructure and properties characterization of a new ceramic filter" *Mater. Design,* 15, 135-140 (1994).

[2] D. L. Trimm and A. Stanislaus, "The control of pore size in alumina catalyst supports " *Appl. Catal.,* **21**, 215-238(1986).

[3] K. Keizer and AJ Burggraaf, "Porous Ceramic Materials in Membrane Applications," *Sci. Ceram.,* **39**, 213-221 (1998).

[4] T. Fukasawa and H. Goto : Japanese patent no.3124274.

[5] T. Fukazawa, M. Ando, T. Ohji, and S. Kanzaki, "Synthesis of Porous Ceramics with Complex Pore Structure by Freeze-Dry Processing " *J. Am. Ceram. Soc.,* **84**, 230-232 (2001).

[6] T. Fukasawa , Z.-Y. Deng, M. Ando1, T. Ohji and Y. Goto," Pore structure of porous ceramics synthesized from water-based slurry by freeze-dry process" *Journal of material science* **36**, 2523-2527 (2001).

NEW PRODUCTION APPROACHES FOR LARGE AND VERY COMPLEX SHAPES OF SILICON NITRIDE AND SILICON CARBIDE CERAMICS

Dr. Karl E. Berroth, FCT Ingenieurkeramik GmbH
Gewerbepark 11, 96528 Rauenstein, Germany
T: +49 36766 868-12, F: +49 36766 868-68, k.berroth@fct-keramik.de

INTRODUCTION

Due to their very specific sets of material properties, silicon nitride and silicon carbide based ceramics have gained a lot of interest in the past 20 years. Many new approaches in technical equipment like gas turbine engines, motor engines, cooking systems, bearings, cutting tools, substrates and ... and many more were investigated with corresponding national and international research activities.

However, except of bearing parts and cutting tool tips, burner nozzles and armors none of these expected high volume applications yet went into industrial production.
On the other side, many less spectacular applications are, as "spin off" of this research state of the art today and have opened a wide field for niche products and led to new technical solutions with higher service time, less wear and corrosion and improved process and product properties. Today, besides the materials technology, the more important requirement is the economic and reliable production of even large and very complex shaped components with very narrow tolerances in µm range.

Therefore at FCT Ingenieurkeramik GmbH large efforts were made during the last years, to get into a position to offer even very large and complex shaped components made of sintered silicon carbide (SSiC) and of gas pressure sintered silicon nitride (GPSN) ceramics. This was reached by developing highly efficient and reliable production procedures and investment in very specific and unique equipment. This so created availability has opened new fields of application for our ceramic material and new markets for us as a ceramic producer. On the other side, designers and engineers are now allowed to think much more complex in designing of ceramic components as in the past. This opens additional technical opportunities.

In the paper, very complex components and the newly developed fabrication routines with rapid prototyping approaches and different kinds of final machining procedures for them. Also the set of properties of the corresponding materials are highly valuable and required for innovative optical applications, for dynamic materials testing, for electronic test equipment for liquid metal processing, metal forming, mechanical, chemical and high temperature engineering will be presented.

Higher resolution due to a very low coefficient of thermal expansion, higher stiffness and a lightweight structure was reached by producing the housing structure of an infrared camera for aerospace with silicon nitride. By using silicon nitride, dynamic testing equipment can reach much higher frequency and so reduce fatigue testing time and costs. With silicon nitride rollers, the lifetime of components in steel rolling mills was extended by more than a magnitude also improving the surface quality of the rolled products. Due to it's high wear resistance and excellent thermal conductivity, our Silicon carbide is highly used as liner and agitator and as cylinder roller for calendering processes in mechanical engineering. There dimensions in the range of m are realized in advanced high tech equipment. Another application are special kiln furniture. Here, the corrosion resistance and the high durability was used, to improve the furnace capacity by a factor of three.

1. MATERIALS AND FABRICATION

FCT Ingenieurkeramik GmbH has established different material grades for gas pressure sintered and hot pressed Silicon nitride (GPSN, HPSN) and sintered silicon carbide (SSiC) as well as for some composite Materials like $C_f/CSiC$ and corresponding processing routes for the commercial fabrication of a broad range of components with mainly large and complex sizes and shapes and i.g. very narrow tolerances.

The material grades are tailored to high durability for specific applications. Parameters are at least on the same level, more likely better as products of our competitors.
Rather large components with one dimension up to 1,25 m but also tiny things with only some mm feed into our product range. We produce prototypes according to customers drawing and also small and intermediate series up to 1000 pieces per year or fabrication lot.

As major fabrication routines we use:
1. slip casting for complex, thin walled components
2. cold isostatic pressing of preforms with subsequent green machining,
3. sintering, hot pressing and hot isostatic pressing
   and if necessary
4. final machining by grinding honing, drilling and polishing.

A typical production routine is shown in the first picture

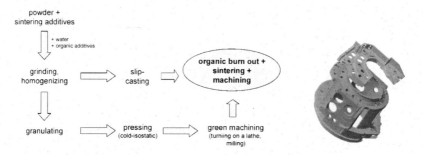

picture 1 production sketch
image

picture 2 3D-CAD

Besides the standard procedures we also can do injection moulding, pressure casting and uniaxial pressing.

As a very specific detail of our capabilities, we can do rapid prototyping by using CAD-CAM technologies, also for very complex and high precision as fired parts. In picture 2, an example of a 3D structure is shown, which is expanded by our shrinkage and wharpage and with grind stock at the spaces where final machining after firing is required.

Another thing is that we can use very different final machining procedures like 3D grinding, laser cutting and drilling as well as spark and wire erosion for electrical conductive grades of silicon carbide and nitride composites.

2. APPLICATIONS
The following examples shall further demonstrate our know how in this aerea.

2.1. Metal working
Rolling

In metal industry, new shaping approaches for even complex parts are rolling operations. For such shaping processes steel, WC and ceramic rollers are used. Silicon nitride due to its excellent thermal shock behaviour, high strength, hardness and toughness shows improved service time and product quality even at rolling temperatures up to 1000°C in rolling of stainless steel. Another advantage is the possibility of cooling reduction and the lack of material transfer to the rolled metal.

But also the use of silicon nitride in guiding elements leads to significant improvements in service time and surface quality. With silicon nitride guiding rolls, the service time for stainless steal wire production was extended from 30 t to up to 2000 t before a regrinding operation was necessary. In total, service loads of more than 15.000 t were reached with reground rollers.

picture 4 press roller barrel   picture 5 foil rolling   picture 6 rollers with silicon nitride roller

Welding

For the production of welded tubes and profiles, highly wear resistant and tough calibration rollers have become state of the art because of their very much extended service time and accuracy.

Also for the fixation and precise, reproducible alignment of steel sheets, fixation pins are widely used in automotive industry. The advantage of silicon nitride in welding is the improved strength, thermal shock capability and the non wetting for sparks created through the welding process.

picture 7 welding rolls and fixation pins

2.3. Mechanical engineering

In mechanical engineering, many applications need a wear resistant material which also can withstand a certain impact stress or mechanical load. Agitator arms and barrels as well as lining inserts in mixers and attrition mills are typical examples for this. They are mainly made in SSiC but also, with higher complexity, GPSN is used too.

picture 8 agitator arm   picture 9   agitator rotor

2.4. Electrical engineering and electronics

Also in electrical engineering, silicon nitride components have gained specific application niches. Highly strong, thermo- mechanical and thermal shock resistant insulator components show improved service behaviour compared to alumina or porcelain. Coil supports for inductive heating equipment for metal heat treatment have tripled the service time.
In electronics, high stiffness and low wear is reached for guiding beams and supporting plates in chip manufacturing equipment.

Also very thin plates of silicon nitride, with diameters up to 385 mm are used for testing devices in the chip production.

Thousands of tiny holes are laser drilled in very specific pattern. Some of them are piled up and afterwards thousands of electrical contact tips were introduced through the pile of sheets, so forming the test device for testing the chips on a new 300 mm wafer in one shot.

2.4. Aluminium foundry industry

Another large application area was developed in Al-casting, two materials properties of silicon nitride are required there:
1. excellent corrosion resistance against metal melt
2. excellent thermal shock resistance
in addition with high strength and fracture toughness which assure a save handling in the tough foundry environment.

Components like thermocouple sheaths, heater sheaths, riser tubes, valve seats and plungers, degassing agitators and a lot of other very specific parts are in use or are tested in new advanced casting systems.

Silicon nitride components are useful for the whole production process of Aluminium parts.

Melting

   Silicon nitride crucibles of can be used for initial melting of metal. Thermal shock, corrosion resistance and high strength guarantee a long service time. Thin walled ladles for taking probes for the inspection of chemical composition as well as for melt dosing are made by slip casting.

 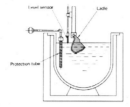 

picture 10 melt crucible, holding furnace with ladle and thermocouple, ladles for melt probing

Melt processing

Thermocouple protection sheaths

   Thermocouple protection sheaths are meanwhile state of the art in temperature measuring systems for aluminium melt. They are produced in three different standards and length up to 1.5 m. Also larger length were realized by connecting two individual tubes by a high temperature mortar.

Immersion heater

   Melt must be processed, in order to have the desired properties for casting. Therefore it must be kept on the right temperature. This can be done by electrically heated or gas fired immersion heaters. Immersion heater sheaths made of gas pressure sintered silicon nitride show service times up to years. The service time is however dependent on proper handling. Immersion heater tubes are used in filter boxes, casting crucibles and as a new approach in liquid aluminium transport containers. In this new application a new usage of liquid metal will be possible also for small foundries. They can keep the melt liquid over several days and can reorganize their production by no longer preparing their own alloy in small portions.

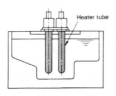

<u>picture 11</u> immersion heater green bodies for a holding furnace, for heating a filter box and for transport aluminium

Degassing agitators

In order to clean the melt, degassing agitators are widely used. They are mainly made of graphite material causing the problem of low service time. Due to the fact, that air and with this oxygen can access to the rotor and more likely to the shaft, the material burns away, and the shaft breaks and more often as a following failure the impeller is destroyed too.

Using GPSN as shaft in an initial step, the service time can be extended from days to months or even years. Using GPSN for rotors is difficult, because only simple geometries are recommended and the costs are very high compared to graphite. If GPSN rotors are done with an adequate design and are properly handled during processing in the foundry, also service times of several months or even years are experienced.

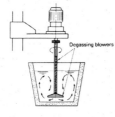

<u>picture 12</u> degassing agitator impeller - large shaft and small lab impeller

Low pressure casting

Riser stalks

For low pressure casting of cylinder heads and wheels a special technology has become state of the art within the last couple years. A melt pot with a removable top which can be pressurized up to 10 bar is typically used. The top has one or two riser stalks who feed the molten metal without contaminations via a transfer tube into the mould. The melt therefore is going up and down within the stalk and so introducing a temperature change which causes cyclic stress. Also corrosion takes place because of the additives which are used for the conditioning of the melt. Depending on size, flange geometry, casting conditions and melt composition, service times from 6 month to 2 years are state of the art. These service times lead to dramatically decreased costs for tubes and handling. A factor of 2 can be considered as a typical value.

Another benefit is the fact, that the downtime for equipment can be reduced from a weekly cycle, which is necessary for cast iron tubes to a six month or even longer cycle for silicon nitride. Additionally nearly no BN coating is required.

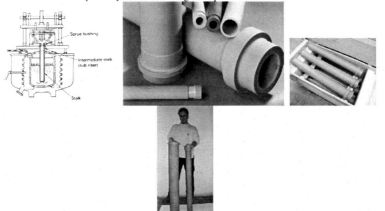

picture 13 low pressure casting equipment for wheels riser tubes of different size and shape

2.6. high temperature engineering

In many high temperature processes like heat treatment of metal parts and sintering processes for technical ceramics, very special shapes of kiln furniture was developed and introduced, using advanced ceramic material. Main requirement is corrosion resistance against high temperature atmosphere and no reaction with the heat treated material.
Due to its very high strength, even at temperature up to 2000°C, SSiC is used as setter plates with a dramatically improved lad capacity for sintering of electronic ceramic devices. The furnace capacity could be tripeled and the also service time for the setters is significantly extended. Picture 16 shows a set of cake shaped setters with position bores for individual ceramic parts.

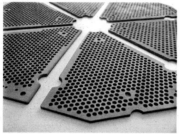

picture 14 setter plates of SSiC

2.1. Optical engineering

For optical applications, mainly for instruments which are used in airplanes or space, the density is a major issue beside stiffness, coefficient of thermal expansion and long term durability. Also mechanical strength and fracture toughness is important, because on starting and landing, very high mechanical load appears. For this large and highly complex components we developed a very sophisticated green machining and sintering technology. Such ceramic structures are required for advanced optical equipment like high resolution telescopes and cameras and have opened a new promising application. Also long term stable guiding beams for optical measuring systems and supporting structures for lenses, can be made from silicon nitride and silicon carbide.

One of the most sensitive parameter for large, structural, monolithic optical components is the coefficient of thermal expansion (CTE). At room temperature, for silicon nitride a value of 1×10^{-6} /K is the lowest available value in combination with low density, high stiffness, strength and durability. The CTE's of corresponding materials for optical use are shown in diagram 1 for the temperature range between -100°C and +100°C. Corresponding components are shown in picture 15

picture 15 camera housing structure and instrumentation platform

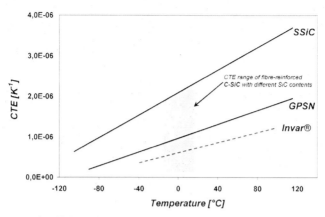

diagram 1 Coefficient of thermal expansion CTE for structural "optical" materials

3. IMPROVED PROPERTIES

As mentioned above, silicon nitride shows improved application benefits compared to other ceramics or metallic solutions. Mainly in wear and foundry industry, solid silicon nitride components gives a much longer service time compared to solutions with coatings. Even more improved properties can be reached by silicon nitride based composite materials.
Here, a wide variety of different additives help, to tailor materials for specific use in new advanced technologies.

With the addition of TiN for example, the hardness and the electrical conductivity can be adjusted to a wide range of technical requirements. Oxide and nonoxide sintering addititives can help to change and improve the corrosion resistance by the reduction of wetability for metal melts. Also the thermal conductivity can be influenced by special rare earth oxides. A wide range of possible compositions is still under investigation.

4. CONCLUSION

In the paper it is shown, that a wide range of applications of silicon nitride ceramics have been achieved due to the availability of components with different sizes, shapes and complexity. Small and very precise as well as rather large and complex components can be reproducibly and economically produced with a high standard of quality. Materials and components are highly reliable, also in very though applications. New technical solutions are generated and though possible for advanced processing equipment and more sophisticated fabrication routines.

FABRICATION AND MECHANICAL PROPERTIES OF POROUS SILICON NITRIDE MATERIALS

Tatsuki Ohji
National Institute of Advanced Industrial Science and Technology (AIST)
Nagoya 463-8560, Japan

ABSTRACT

In structural materials, pores are generally believed to deteriorate the mechanical reliability. This study, however, demonstrates pores can cause improved or unique performance when the sizes, shapes, and orientation of pores as well as grains are controlled, taking an example of porous silicon nitrides. Two types of porous materials are dealt with; one is isotropic porous silicon nitrides, where fibrous grains are randomly oriented, and the other is anisotropic porous silicon nitrides, where fibrous grains are substantially aligned in one direction. Particular emphasis is placed in substantially improved properties including fracture energy, fracture toughness, strain tolerance, and thermal shock resistance of the porous materials, even compared with those of the dense materials.

INTRODUCTION

Porous ceramic materials have been industrialized into permeable applications such as gas filters, separation membranes, and catalyst supports. For such applications, a high porosity (typically >40%) of channels, with a well-controlled pore size distribution, is required. On the other hand, porous ceramic materials with a tailored microstructure have potential to be used as structural components due to unique properties such as low density, good strain and damage tolerance, and good thermal shock resistance.

During the past two decades a lot of researchers have been devoted to controlling the size, shape, morphology and distribution of the grains in ceramics, in order to improve the mechanical properties or to realize outstanding compatibility of competitive properties. Silicon nitride ceramics is one of the most well developed materials through such approaches. For example, both the strength and fracture toughness of silicon nitride ceramics, which are known as antagonistic properties, can be improved concurrently, by controlling morphologies and alignment of the fibrous grains [1-4]

In structural materials, pores are generally believed to deteriorate the mechanical reliability; however, this is not always true whenever the microstructure is carefully controlled. The contents of this review show how the mechanical properties change when the size, shapes, and orientation of pores as well as grains are controlled, taking an example of porous silicon nitrides. Particular emphasis is placed in unique or improved properties including fracture energy, fracture toughness, strain tolerance, and thermal shock resistance through such porous structure-control. The contents are roughly divided into two parts; one is for isotropic porous silicon nitrides, where fibrous grains are randomly oriented, and the other is for anisotropic porous silicon nitrides, where fibrous grains are substantially aligned in one direction.

ISOTROPIC POROUS SILICON NITRIDES
Microstructure Development and Mechanical Properties

Several processes have been developed to produce the porous silicon nitride ceramics. Sintering powder compacts to a fixed degree of densification, a process called partial sintering is one of the most frequently employed approaches to fabricate porous ceramic materials. In the case of oxide materials, due to their good sinterability, the density or porosity can be adjusted by the heating characteristics, such as temperature and holding time. The sintering of silicon nitride ceramics is difficult, however, because of strong covalent bonding between silicon and nitrogen atoms, so that sintering additives are necessary for consolidating silicon nitride ceramics by liquid-phase sintering. On the other hand, the difficulty of sintering silicon nitride ceramics is beneficial for controlling density or porosity through adjusting the additives and the sintering process. A fibrous silicon nitride microstructure is developed during sintering at adequately high temperatures in porous material, and excellent mechanical properties can be obtained [5, 6], compared with those of porous oxide ceramics obtained by partial sintering.

Yang, *et al.* [7] fabricated porous silicon nitride ceramics using Yb_2O_3 sintering additives and investigated the microstructures and mechanical properties of those ceramics, as a function of porosity. Sintering was performed, at various temperatures between 1600 and 1850°C, under a nitrogen-gas pressure of 0.6 MPa. A liquid-phase sintering technique and powder compaction similar to the fabrication of dense silicon nitride were used to make the fabrication process simple and cost-effective. The sintering additive, Yb_2O_3, is known to form crystalline $Yb_4Si_2O_7N_2$ at grain boundaries or triple junctions [8, 9] which presumably enhances the high-temperature mechanical properties of silicon nitride ceramics. The addition of Yb_2O_3 also is known to accelerate the fibrous grain growth of β-Si_3N_4 [10], which is advantageous for the

Figure 1. Microstructures of porous silicon nitride ceramic with 5 wt.% Yb_2O_3 sintered at 1600°C, 1700°C, 1800°C and 1850°C, respectively. "P" denotes porosity. Reprinted from [7], with permission from Elsevier Ltd. All rights reserved.

crack shielding effects of grain bridging and pullout. Furthermore, because of the high melting point of Yb_2O_3 and the high viscosity of Yb_2O_3-related glass, this additive is unsuitable for the densification of silicon nitride, a characteristic that is, conversely, beneficial for the fabrication of porous silicon nitride.

The microstructures of the resultant silicon nitride porous ceramics after sintering at different temperatures are shown in Fig. 1. The microstructure of the sample sintered at 1600°C consisted solely of fine, equiaxed grains. The average grain size was almost the same as that of the starting powder, indicating little phase transformation, and the XRD analysis identified almost only α-Si_3N_4 peaks. When the sintering temperature increased above 1700°C however, the formation and development of β-Si_3N_4 grains were observed, indicating enhanced phase transformation and grain growth. At 1700°C, very fine, fibrous β-Si_3N_4 grains were obtained, and with increasing the sintering temperature, the microstructure becomes coarse while the porosity changes relatively little.

The fracture strength, σ_f, fracture energy, γ, and fracture toughness (critical stress intensity factor), K_{Ic} are shown in Fig. 2, as a function of sintering temperature. Each data point marks an average of five or six measurements. The fracture strength and fracture energy were determined by three-point flexure [11] and chevron-notched beam (CNB) [12] tests, respectively, and the fracture toughness was converted from the fracture energy using the following relationship.

$$\gamma_{eff} = K_{Ic}^2 \, (1-v^2) \, / \, 2E \tag{1}$$

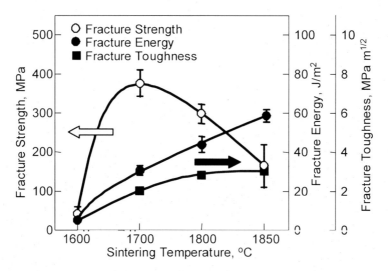

Figure 2. Mechanical properties of porous silicon nitride ceramics with 5 wt.% Yb_2O_3 sintered at 1600°C, 1700°C, 1800°C and 1850°C, whose microstructures are shown in Fig.1. Note that porosities of materials sintered above 1700°C do not change very much.

where E is the Young's modulus, and ν is the Poisson's ratio. The reported fracture toughness is an averaged value when the toughness varies with crack extension (R-curve). When the sintering temperature is 1600°C and the microstructure is equiaxed, the mechanical properties are low; however, the properties are markedly improved along with the fibrous grain formation above 1700°C. For example the sample sintered at 1700°C has a strength of 380 MPa, which is about 10 times larger than that at 1600°C. As the sintering temperature further increased and the microstructure became coarser, the strength decreased while the fracture energy and converted toughness increased. The toughness values are still lower than 4 MPa m$^{1/2}$.

Porosity Dependence of Mechanical Properties

The effects of porosities on the mechanical properties of porous silicon nitride ceramics have also been studied; the porosity was precisely controlled from 0 to 30% by using partial hot-pressing (PHP). In the PHP, the porosity was controlled by the configuration of the carbon mold and the powder amount, while leaving other parameters such as characteristics of a powder mixture and sintering additives fixed. They used 5 wt% Yb$_2$O$_3$ as the sintering additive, and sintering was conducted at 1800°C using nitrogen atmosphere for two hours by using a specially designed mold where the sum of top and bottom punch lengths is shorter than the length of the mold [5].

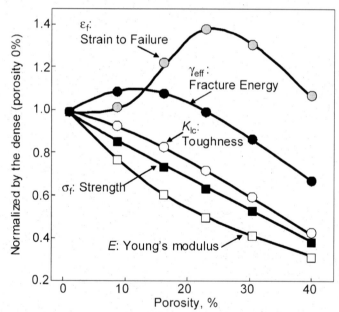

Figure 3. Young's modulus, E, fracture strength, σ_f, fracture energy, γ_{eff}, fracture toughness, K_{IC}, and failure to strain, or strain tolerance, ε_f, (obtained from the E and σ_f). The values are normalized by the respective values of the dense material; $E_0 = 330$ GPa, $K_{IC0} = 6.3$ MPa m$^{1/2}$, $\gamma_{eff0} = 70$ J/m^2 and $\varepsilon_{f0} = 0.0034$.

Mechanical properties including Young's modulus, E, fracture strength, σ_f, fracture energy, γ_{eff}, and fracture toughness, K_{IC}, as a function of porosity are shown in Fig. 3 [6]. The values are normalized by the respective values of the dense material: $\sigma_{f0} = 1.1$, GPa, $E_0 = 330$ GPa, $K_{IC0} = 6.3$ MPa m$^{1/2}$, $\gamma_{eff0} = 70$ J/m^2 and $\varepsilon_{f0} = 0.0034$. It should be noted that the fracture energy first increased somewhat with increasing the porosity up to 10 – 20%, and then decreased while all the other properties continuously decreased. The degree of decrease in the Young's modulus is larger than that in the strength in the porosity range of 20 – 30%; therefore the failure to strain, or strain tolerance, ε_f, given by the following equation, has the maximum in this range (Fig. 3).

$$\varepsilon_f = \sigma_f / E \qquad (2)$$

The existence of pores lowers the strength, Young's modulus and fracture toughness, but increases the facture energy and strain tolerance depending on the porosity, in the isotropic porous silicon nitrides.

ANISOTROPIC POROUS SILICON NITRIDES
Fabrication and Microstructures

There have been many studies for toughening silicon nitride ceramics by controlling microstructures, such as grain size and morphology, grain alignment, and boundary chemistry to enhance a variety of crack wake toughening mechanisms including grain bridging and grain pull-out. [1-4] Aligning fibrous silicon nitride grains using a tape-casting technique with seed crystals, Hirao et al. [3] obtained high fracture toughness and high fracture strength, when a stress was applied parallel to, or a crack extended normal to, the grain alignment. In this case, a greater number of fibrous grains are involved with the crack wake toughening mechanisms, and then the toughness can rise steeply in a very short crack extension [4].

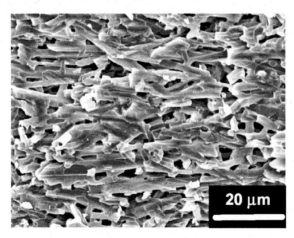

Figure 4. Microstructures of anisotropic porous silicon nitride prepared by tape-casting fibrous seed crystals (porosity: 14%). Reprinted from [11], with permission from Blackwell Publishing, Inc. All rights reserved.

Taking a similar approach, Inagaki *et al.* [13, 14] developed a porous silicon nitride, where the fibrous grains were uniaxially aligned, and investigated the effects of pores on the mechanical properties. Since the pores exist around the aligned fibrous grains, enhanced operations of the grain bridging and pull-out can be anticipated when the crack propagates perpendicularly to the direction of grain alignment. Furthermore, the presence of pores surrounding the silicon nitride grains causes cracks to tilt or twist, namely crack deflections [15, 16]. As starting materials they used β-Si$_3$N$_4$ seed crystals, which were obtained by heating the powder mixture of α-Si$_3$N$_4$, 5 mol.% Y$_2$O$_3$, and 10 mol.% SiO$_2$ in a silicon nitride crucible at 1850°C for 2 h [17]. The slurry containing the fibrous seed crystals mixed with 5 wt% Y$_2$O$_3$ and 2 wt % Al$_2$O$_3$ was tape-cast so that the seed crystals in the sheets were aligned along the casting direction. After the green sheets were stacked and bonded under pressure, sintering was performed at 1850°C under a nitrogen pressure of 1 MPa. The texture of porous silicon nitride with porosity of 14% was shown in Fig. 4; the material consists only of fibrous grains, pores and grain boundary phase. The fibrous grains of silicon nitride tend to be aligned along the casting direction, but many of the grains are substantially tilted. The pores, whose shapes are mostly plate-like along the same direction, exist among the grains.

Porosity Dependence of Mechanical Properties

The effect of porosity on the mechanical properties of the anisotoropic porous silicon nitride is investigated in both directions parallel to and perpendicular to the grain alignment.

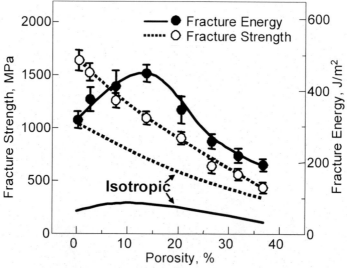

Figure 5. Porosity dependence of fracture strength and fracture energy of anisotropic porous silicon nitrides, in comparison with those of the isotropic materials shown in Fig. 3. Stress is applied parallel to the alignment direction. Bar is the standard deviation.

Figure 5 shows the porosity dependencies of fracture strength and fracture energy when the stress is applied in the parallel direction to the grain alignment of the porous silicon nitride, in comparison with those of the isotropic materials shown in Fig. 3. The strength was measured by the three-point flexural test while the fracture energy was determined by the CNB technique. The numbers of measurements for the former and latter are six and three, respectively, and the plots in the figure are their averages. The fracture strength became larger as the porosity decreased. In the porosity range below 5%, the strength attained above 1.5 GPa. This value was almost comparable to that of fibrous-grain-aligned dense silicon nitride fabricated through superplastic forging [18]. When fibrous grains were aligned, pores around the grains promoted debonding between interlocking fibrous grains without broken. Therefore, even small amount of pores enhanced the grain bridging improving the strength. On the other hand, fracture energy of porous silicon nitride ranged from 300 to 500 J/m^2 in the porosity range below 20%. These values were considerably high compared with other studies [12, 19]. Large fracture energy was mainly due to the effect of bridging crack by aligned fibrous grains and/or pullout of the grains [13]. Caused to debond by the existence of pores, aligned grains bridging the crack or interlocking each other were drawn apart without breaking, which resulted in increase of sliding resistance. As porosity increased to nearly 15%, fracture energy became larger, however, fracture energy decreased monotonically in the porosity range over 15% with increase of the porosity. This drop of fracture energy is presumably due to the reduction of substantial bridging and/or pullout area.

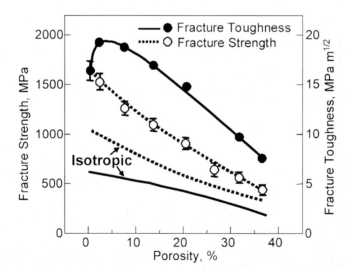

Figure 6. Porosity dependence of fracture strength and fracture toughness, K_{IC} of of anisotropic porous silicon nitrides, in comparison with those of the isotropic materials shown in Fig. 3. Stress is applied parallel to the alignment direction. Bar is the standard deviation.

The porosity dependences of the fracture toughness estimated from the fracture energy using the equation 1 as well as the fracture strength are shown in Fig. 6, in comparison with those of the isotropic materials. High fracture toughness above 17 MPa m$^{1/2}$ as well as high strength above 1.5 GPa was attained in the porosity range below 5%. Interestingly, the fracture toughness of specimens with very small amounts of porosity (<5%) exceeded that of fully dense materials, and approached 20 MPa m$^{1/2}$. The fracture toughness shown here is an averaged value over the toughness varying with crack extension, as stated earlier. Comparing the load-displacement curves of CNB tests for the porous silicon nitrides with 4% and 14% porosities, the maximum load-point of the former is higher than that of the latter, suggesting the steep rise of fracture resistance with a short crack extension. Similar rapid rise in the R-curve has been reported for alumina and other dense silicon nitrides [20].

Thermal Shock Resistances
Generally mechanical strength of ceramics after thermal shock remains constant until the temperature difference, ΔT, reaches a critical value, ΔT_c which is determined by the shape and sizes of specimen and heat transfer conditions, in addition to the material properties. When $\Delta T > \Delta T_c$, a crack propagates catastrophically or quasi-statically depending on the initial crack size, leading to an abrupt or gradual decrease of strength. Therefore, two parameters, "thermal shock *fracture* resistance", R, and "thermal shock *damage* resistance", R'''', are often used to characterize the resistance against thermal shock crack initiation and propagation, respectively. As defined by Hasselman [21, 22], R and R'''' given as

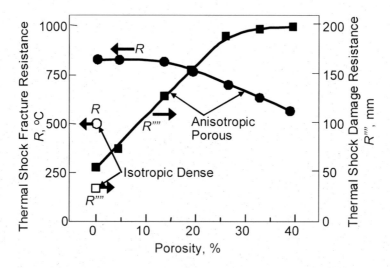

Figure7. Estimated thermal shock fracture resistance, R, and thermal shock damage resistance, R'''', of the anisotropic porous silicon nitride as a function of porosity, in comparison with those of dense fine-grained material.

$$R = \sigma_f (1 - \nu) / E\alpha \ \Delta T_c \qquad\qquad (3)$$

$$R'''' = E \ \gamma_{eff} / \sigma_f^2 (1 - \nu) \qquad\qquad (4)$$

where α is the coefficient of thermal expansion. It should be noted that the effects of fracture strength and Young's modulus on R and R'''' are completely opposite; when the fracture strength is high and the Young's modulus is low, R is high but R'''' is low. From the properties including the fracture strength, Young's modulus, and fracture energy, it is possible to estimate R and R'''' of the anisotropic porous silicon nitride as a function of porosity. Figure 7 shows the estimated R and R'''' in comparison with those estimated for the dense, isotropic material whose properties are 1.1 GPa, 330 GPa, 0.25 and 70 J/m^2 for the fracture strength, Young's modulus, Poisson's ratio and fracture energy, respectively. The thermal shock fracture resistance, R, remains high in the porosity region up to 20% and then starts to decrease. On the other hand, the thermal shock damage resistance, R'''', increases with increasing porosity up to 30% followed by the plateau region afterwards. It should be noted that high values of both R and R'''' can be realized in the porosity range of 15~25%, which are much higher than those of the dense, isotropic material.

In order to experimentally characterize R, of the anisotropic porous silicon nitride (14% porosity) in comparison with the above dense material, thermal shock tests were performed by dropping the heated specimens from a resistance furnace into a container of water at 20°C [23]. The specimen was covered with mullite blocks on all sides except one face exposed to the water quenching. After heating in air at a rate of 5°C/min to a preset temperature and holding at this temperature for 30 min, the specimen was quenched into a water bath at 20°C. The fracture

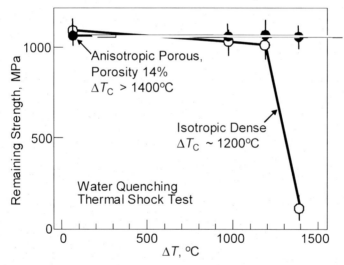

Figure 8. Flexural strength as a function of quenching-temperature difference for anisotropic porous and isotropic dense silicon nitrides.

strengths of the quenched specimens were determined at room temperature by a three-point bending test with a support distance of 30 mm and a crosshead speed of 0.5 mm/min, where the water-exposed side was used as the tensile surface. The results are shown in Fig. 8; the anisotropic porous material shows a thermal shock fracture resistance whose ΔT_c is larger than 1400°C, while that of the dense is about 1200°C.

SUMMARY

In structural materials, pores are generally believed to deteriorate the mechanical reliability. This study, however, demonstrated pores could cause improved or unique performance when the sizes, shapes, and orientation of pores as well as grains were controlled in the porous silicon nitrides. As for the isotropic porous materials where fibrous grains are randomly oriented, the existence of pores lowers the strength, Young's modulus and fracture toughness, but increases the facture energy and strain tolerance depending on the porosity. As for the anisotropic porous materials where fibrous grains are substantially aligned, high fracture energy from 300 to 500 J/m^2 was obtained in the porosity range below 20%, and very high fracture toughness above 17 MPa $m^{1/2}$ as well as high strength above 1.5 GPa were attained below 5%. Substantial improvements in both the thermal shock fracture resistance, R, and thermal shock damage resistance, R'''', were also predicted, even compared with those of the dense fine-grained material. The improvement in R was confirmed by the water quenching thermal shock test.

REFERENCES

[1]P. F. Becher, E. Y. Sun, K. P. Plucknett, K. B. Alexander, C. H. Hsueh, H. T. Lin, S. B. Waters, C. G. Westmoreland, E. S. Kang, K. Hirao and M. E. Brito, "Microstructural Design of Silicon Nitride with Improved Fracture Toughness: I, Effects of Grain Shape and Size," *J. Am. Ceram. Soc.*, **81**, 2821–2830 (1998).

[2]E. Y. Sun, P. F. Becher, K. P. Plucknett, C. H. Hsueh, K. B. Alexander, S. B. Waters, C. G. Westmoreland, K. Hirao and M. E. Brito, "Microstructural Design of Silicon Nitride with Improved Fracture Toughness: II, Effect of Additives," *J. Am. Ceram. Soc.*, **81**, 2831–2840 (1998).

[3]K. Hirao, M. Ohashi, M. E. Brito, and S. Kanzaki, "A Processing Strategy for Producing Highly Anisotropic Silicon Nitride," *J. Am. Ceram. Soc.*, **78**, 1687–1690 (1995).

[4]T. Ohji, K. Hirao, and S. Kanzaki, "Fracture Resistance Behavior of Highly Anisotropic Silicon Nitride," *J. Am. Ceram. Soc.*, **78**, 3125–3128 (1995).

[5]J. F. Yang, G. J. Zhang, and T. Ohji, "Porosity and Microstructure Control of Porous Ceramics by Partial Hot-Pressing," *J. Mater. Res.*, **16**, 1916–1918 (2001).

[6]J. F. Yang, G. J. Zhang, and T. Ohji, "Microstructure and Mechanical Properties of Silicon Nitride Ceramics with Controlled Porosity," *J. Am. Ceram. Soc.*, **84**, 1639–1641 (2001).

[7] J. F. Yang, Z. Y. Deng, and T. Ohji, "Fabrication and Mechanical Properties of Porous Si_3N_4 Ceramics Doped with Yb_2O_3 Additions," *J. Euro. Ceram. Soc.*, **23**, 371–378 (2003).

[8] C. M. Wang, X. Pan, M. J. Hoffmann, R. M. Cannon and M. Ruhle, "Grain Boundary Films in Rare-Earth-Glass-Based Silicon Nitride," *J. Am. Ceram. Soc.*, **79**, 788–792 (1996).

[9] T. Nishimura and M. Mitomo, "Phase relationship in the system Si_3N_4-SiO_2-Yb_2O_3," *J. Mater. Res.*, **10**, 240–242 (1995).

[10]H. J. Park, H. E. Kim, and K. Niihara, "Microstructural Evaluation and Mechanical Properties of Si_3N_4 with Yb_2O_3 as a Sintering Additive," *J. Am. Ceram. Soc.*, **80**, 750–756 (1997).

[11] JIS R 1601:1995, "Testing method for flexural strength (modulus of rupture) of fine ceramics."

[12] T. Ohji, Y. Goto, and A. Tsuge, "High-Temperature Toughness and Tensile Strength of Whisker-Reinforced Silicon Nitride," *J. Am. Ceram. Soc.*, **74**, 739–745 (1991).

[13] Y. Inagaki, T. Ohji, S. Kanzaki, and Y. Shigegaki, "Fracture Energy of an Aligned Porous Silicon Nitride," *J. Am. Ceram. Soc.*, **83**, 1807-1809 (2000).

[14] Y. Inagaki, Y. Shigegaki, M. Ando, and T. Ohji, "Synthesis and Evaluation of Anisotropic Porous Silicon Nitride," *J. Eur. Ceram. Soc.*, **24**, 197-200 (2004).

[15] K. T. Faber and A. G. Evans, "Crack Deflection Processes I, Theory," *Acta. Metall.*, **31**, 565–576 (1983).

[16] K. T. Faber and A. G. Evans, "Crack Deflection Processes II. Experimental," *Acta. Metall.*, **31**, 577–584 (1983).

[17] Y. Inagaki, M. Ando, and T. Ohji, "Synthesis of Porous Silicon Nitride by Adding Seed Crystals," *J. Ceram. Soc. Japan*, **109**, 978-980 (2001).

[18] N. Kondo, T. Ohji, and F. Wakai, "Strengthening and Toughening of Silicon Nitride by Superplastic Deformation," *J. Am. Ceram. Soc.*, **81**, 713–716 (1998).

[19] N. Kondo, Y. Inagaki, Y. Suzuki, and T. Ohji, "High-Temperature Fracture Energy of Superplastically Forged Silicon Nitride," *J. Am. Ceram. Soc.*, **84**, 1791–1796 (2001).

[20] J. A. Salem, G. D. Quinn, and M. G. Jenkins, "Measuring the Real Fracture toughness of Ceramics –ASTM C1421", pp. 531-54 in *Fracture Mechanics of Ceramics*, Vol. 14. Edited by R.C. Bradt, D. Munz, M. Sakai, and K.W. White. Springer, New York, NY, 2005

[21] D. P. H. Hasselman, "Unified Theory of Thermal Shock Fracture Initiation and Crack Propagation in Brittle Ceramics," *J. Am. Ceram. Soc.*, **52**, 600-604 (1969).

[22] D. P. H. Hasselman, "Strength Behavior of Polycrystalline Alumina Subjected to Thermal Shock," *J. Am. Ceram. Soc.*, **53**, 490-495 (1970)

[23] J.-H. She and T. Ohji, unpublished work.

APPLICATION OF TAGUCHI METHOD IN THE OPTIMIZATION OF PROCESS PARAMETERS FOR CONICITY OF HOLES IN ULTRASONIC DRILLING OF ENGINEERING CERAMICS

R. C. S. Mehta[1], R. S. Jadoun[1*], Pradeep Kumar[2], B. K. Mishra[2]

[1]Production Engineering Department, College of Technology, Pantnagar-263 145 (INDIA)
[2] Mechanical and Industrial Engineering Department, IIT Roorkee, Roorkee-247667 (INDIA)

ABSTRACT

This paper presents a study of the effect of process parameters on conicity of holes drilled by USM in alumina based ceramics using silicon carbide (SiC) abrasive. Conicity is defined as the difference between the hole diameters at the entry side and exit side per unit length of the hole produced. It is mainly due to lateral wear of the tool and use of harder tool materials. In order to achieve the objective of the present study, the experiments were conducted on an 'AP-500 model Sonic-Mill' ultrasonic machine.The parameters considered are workpiece material, tool material, grit size of the abrasive, power rating and slurry concentration. Taguchi's optimization approach is used to obtain the optimal parameters. The result shows that the conicity increases almost linearly as the grain size increases. The conicity also increases marginally with the increase in power and concentration. There is a significant interaction between tool & work piece, work piece & grit size and tool & grit size with regard to the raw data. The significant parameters are also identified and their effect on conicity is studied. The results obtained are validated by conducting the confirmation experiments.

INTRODUCTION

The twentieth century has witnessed the creation of products made from the most durable and consequently, the most un-machinable materials in the field of aerospace, nuclear science, automobiles and defense research. Ceramics, Ferrite, Glass, Hastalloy, Nimonics, Piezo-electrics, Quartz, Sapphire, Silicon, and Silicon Carbide etc. are some such materials being used in the above industries [1]. Out of these materials, advanced engineering ceramics have many attractive properties such as high hardness, high thermal resistance, chemical stability, and good insulating aspects.

The conventional manufacturing processes are difficult, time consuming, and sometimes impossible and so, un-economical for harder and difficult to machine materials like ceramics and also give poor accuracy & surface finish. To tackle the problem, the newer machining processes have been developed, which are often called 'modern machining processes' or 'unconventional machining processes' or 'advanced manufacturing processes'. Ultrasonic machining is a valuable process for the precision machining of hard, brittle materials because of many of it's unique characteristics [2]. The tool (shaped conversely to the desired hole or cavity) oscillates at high frequency (typically 20 kHz) and is fed into the work piece by a constant force. Abrasive slurry comprising water and small abrasive particles is supplied between the tool tip and the work piece. Material removal occurs when the abrasive particles, suspended in the slurry between the tool and work piece, impact the work piece due to the down stroke of the vibrating tool.

Quality achieved by means of design optimization is found by many manufacturers to be cost effective in gaining and maintaining a competitive position in the world market [3]. Thus, the main thrust of this investigation is parametric optimization (an off-line quality control activity) with regards to the conicity of holes produced in USM process.

REVIEW OF LITERATURE

Kennedy and Grieve[4] has reported that the factors affecting accuracy of USM are: the precision of the machine tool (i.e. the accuracy of the feed motion), the accuracy of the fixtures used, the quality of the assembly element, abrasive grit size, tool wear, transverse vibration effects, and depth of cut.

The amount of oversize of the holes is greater at entry than at exit resulting in unavoidable conicity due to tool wear. The amount of oversize at the bottom of the hole is of the same order as the smallest abrasive size. Conicity can be reduced by using tungsten carbide and stainless steel tool materials [5], an internal slurry delivery system [5, 6, 7], tools with negative tapering walls or fine abrasives [4, 7, 8, 9]. Dimensional accuracy of the order of ±5 μm can be obtained in most materials. Conicity is reduced at higher static loads and for prolonged operating times since tool wear is less with finer abrasives [5,8]. Use of combined tools with negative taper improves accuracy [10]. Injection of slurry into machining zone increases precision and decreases conicity [10,11] . Re-passing with the use of fine abrasives can eliminate conicity [11,12] .

From the above, it is observed that much of the emphasis is laid on the methods to improve the machining rate and to find out the stress distribution in the work piece. Some of the researchers also made an attempt to study other performance characteristics by varying one factor at a time. Thus, literature lacks in systematic investigation of the effect of process parameters on the quality of drilled hole. Moreover, most of the studies conducted earlier, have not considered the interaction effects of process parameters on quality of drilled hole. However, quality of the drilled hole, which can be estimated from such attributes as the obtainable surface finish, hole oversize and conicity, is dependent on the complex interaction of process parameters like slurry concentration, grit size, power, work piece material etc. Thus, there is a need to further investigate the effect of different process parameters on the quality of hole drilled by ultrasonic machining.

MATERIALS AND METHODS

Means and materials
In order to achieve the objective of the present study, the experiments were conducted on an 'AP-500 model Sonic-Mill' ultrasonic machine. Hot pressed Alumina based ceramic composites were used in this investigation. A monolithic Al_2O_3 was used as the baseline material. SiC particles (average particle size 1 μm) were added to Al_2O_3 matrix according to the combinations listed in Table 1. The Workpieces were cast in the plate form of size 38.1 x 38.1 x 6.35 mm.

Table 1: Composition of ceramic composites[*]

Material Designation	Contains %age of weight			
	Al_2O_3	CaO	SiC	MgO
1	50	25	20	5
2	60	15	20	5
3	70	5	20	5

* Supplied by the manufacturer

The tools made of high carbon steel, high speed steel and tungsten carbide are used in this investigation. The tools were silver brazed to the replaceable threaded tip. Brazing was done at 1200^0 F (648^0 C). Before brazing, the alignment of tool and replaceable threaded tip is ensured and then brazing is done with utmost care, so as to keep the axis of the horn and the axis of the tool in line.

Silicon carbide was used as an abrasive to drill the hole by USD process. Silicon Carbide is a high quality abrasive available in two type's viz. black & green. Due to availability of Black silicon carbide, the abrasive with properties such as hardness on mohs scale (9.7); fracture toughness (4.5 MPa.m$^{1/2}$); specific gravity (3.2 gm / cc); Young's modulus (440 GPa); melting Point (2600 ° C); black color is used in the present investigations. On the basis of pilot experiments the range of the grit size has been decided as # 220, 320 and 500. Water is used as liquid media to make abrasive slurry. The range of the slurry concentration was decided on the basis of literature review and pilot experiments conducted with selected process parameter at different values using one factor at a time approach. The selected concentrations are 25%, 30% and 35 %.

Measurements
The measurements of various properties reported in this paper are done as per the procedure specified in ASTM standards 2001. The increase in the size of the hole produced with reference to the size of the tool is known as the oversize of the hole produced. The diameter of the hole at the entry side was measured by using Tool Maker's Microscope. The tool diameter was subtracted from the hole diameter to get the hole oversize. Out-of-roundness refers to the errors of geometrical form of the circular holes drilled. Diameter of measuring out-of-roundness or circularity is most widely preferred method [13] . For this, diameters at 3 different places were measured using Tool Maker's Microscope. Thus the 'out-of-roundness' was calculated as the difference between highest and lowest diameters of the drilled hole. Conicity is defined as the difference between the hole diameters at the entry side and exit side per unit length of the hole produced.

Plan of experiments (Taguchi's Technique)
This paper uses Taguchi's method, which is very effective to deal with responses influenced by multi-variables. Taguchi's method of experimental design provides a simple, efficient and systematic approach to determine optimal machining parameters. Taguchi recommends orthogonal arrays (OA) for laying out of experiments. For optimum performance characteristics

of the USM process, five process parameters viz. workpiece material (A), tool material (B), grit size (C), power rating (D), slurry concentration (E) and three two-parameter interactions viz. AxB, BxC, AxC were selected as shown in Table 2. Berne and Taguchi [16] have identified that the non-linear behavior (if any) of the parameters of a process can only be determined if more than two levels are used. As per Ross [14] , it is also necessary that the interval between the levels in multi-level experiment must be equal. Hence, it was decided to study each selected parameter at three levels. With five parameters each at three levels and three second order interactions, the total DOF required is 22 [= 5 x (3-1) + 3 x 4], since a three level parameter has 2 DOF (No. of levels − 1) and each two-parameter interaction term has 4 DOF (2x2). Hence, an L_{27} (3^{13}) OA (a standard 3-level OA) has been selected for this phase of experimental work. The L_{27} OA [14] with assignment of parameters and interactions has been used here. The parameters and interactions have been assigned to specific columns of the OA using the Triangular Table [14] and linear graphs [15].

According to the scheme of experimentation outlined in standard L_{27} OA [14], holes were drilled in the workpieces. Three repetitions per trial, i.e. three holes were drilled at every trial condition, resulting in a total of 81 tests. The experimental results are shown in Table 3.

DATA ANALYSIS

The signal to noise ratios is obtained using Taguchi's methodology. Here, the term 'signal' represents the desirable value (mean) and the 'noise' represents the undesirable value (standard deviation). Thus, the S/N ratio represents the amount of variation present in the performance characteristic. Depending upon the objective of the performance characteristic there can be various types of S/N ratios. Here the desirable objective is to obtain lower values of conicity. Hence, the lower-the-better type S/N ratio, as given below was applied.

$$(S/N)_{LB} = - 10 \log \left[\frac{1}{R} \sum_{j=1}^{R} y_j^2 \right]$$

(1)

where y_j = value of the characteristic in an observation j
       R = number of observation or number of repetitions in a trial
The S/N ratios for CC, calculated from the observed data are shown in Table 4.

Table 2: Process parameters and their values at different levels

Process Parameters Symbols	Process Parameter	Level 1	Level 2	Level 3
A	Work Piece	50% Al_2O_3	60% Al_2O3	70% Al_2O_3
B	Tool	HCS	HSS	TC
C	Grit Size	220	320	500
D	Power Rating	40%	50%	60%
E	Slurry Concentration	25%	30%	35%

Constant Parameters	
Frequency of Vibrations	20 kHz
Static Load	1000 g
Type of Work Piece	Alumina based Ceramic
Thickness of Work Piece	5 mm
Tool Geometry	As shown in Figure
Abrasive Type	Silicon Carbide
Liquid Media	Water
Slurry Temperature	20 ° C i.e. near the ambient temperature of tap water
Flow Rate	50×10^3 mm^3 / min

.

Table 3: Experimental results of out-of roundness: raw data and S/N ratios

EXPT.NO	RESPONSES Conocity			AVERAGE= $(R1+R2+R3)/3$ $\bar{y}$	SIGNAL TO NOISE RATIO (dB) (S/N)
	R1	R2	R3		
1	0.048	0.046	0.049	0.048	26.446
2	0.042	0.042	0.042	0.042	27.549
3	0.037	0.037	0.038	0.037	28.542
4	0.040	0.041	0.041	0.041	27.786
5	0.041	0.042	0.040	0.041	27.730
6	0.030	0.030	0.031	0.030	30.341
7	0.039	0.039	0.040	0.039	28.089
8	0.022	0.021	0.022	0.022	33.176
9	0.014	0.014	0.015	0.014	36.872
10	0.057	0.058	0.056	0.057	24.882
11	0.046	0.046	0.047	0.046	26.669
12	0.037	0.036	0.037	0.037	28.651
13	0.050	0.049	0.051	0.050	26.019
14	0.042	0.042	0.042	0.042	27.507
15	0.035	0.035	0.034	0.035	29.218
16	0.039	0.040	0.040	0.039	28.075
17	0.024	0.024	0.024	0.024	32.444
18	0.018	0.016	0.015	0.016	35.713
19	0.067	0.068	0.069	0.068	23.349
20	0.052	0.051	0.050	0.051	25.882
21	0.044	0.045	0.044	0.044	27.091
22	0.052	0.053	0.053	0.053	25.558
23	0.044	0.045	0.045	0.045	27.026
24	0.036	0.037	0.038	0.037	28.651
25	0.064	0.066	0.066	0.065	23.706
26	0.050	0.049	0.049	0.049	26.184
27	0.027	0.026	0.026	0.026	31.655

The lebel average responses from the raw data help in analysing the trend of the quality characteristic with respect to the variation of the factors under study. The level average response plots based on the S/N data help in optimizing the objective under study. The average response plots for raw and S/N data are shown in Figures 1. The interactions effect of the factors under study have also been considered.

Tables 4 show the results of the pooled ANOVA with the conicity (CC) in workpiece. This analysis was carried out for a level of confidence of 95%. The last column of the tables previously mentioned shows the percentage contribution of each factor on the total variation which indicates the degree of influence on the result. The percentage contributions of significant parameters are plotted as shown in Figures 2.

The optimal levels of process parameters are selected on the basis of average response analysis for S/N data and pooled ANOVA. Whatever may be the objective of quality characteristic, the peak points in the S/N response graph for significant parameters give the optimal combination for the quality characteristic. The optimal settings for conicity are given in Table 5.

After determination of the optimum condition, the mean of the response (μ) at the optimum condition is predicted. This mean is estimated only from the significant parameters. The estimate of the mean (μ) is only a point estimate based on the average of results obtained from the experiment. Statistically this provides a 50% chance of the true average being greater than μ and a 50 % chance of true average being less than μ. It is therefore customary to represent the values of a statistical parameter as a range within which it is likely to fall, for a given level of confidence [14].This range is termed as the confidence interval (CI). In other words, the confidence interval is a maximum and minimum value between which the true average should fall at some stated percentage of confidence [14]. The two types of confidence intervals are suggested by Taguchi in regards to the estimated mean of the optimal treatment condition [14]. The expressions for computing these confidence intervals are given below [14, 15].

$$CI_{pop} = \sqrt{\frac{F_\alpha(1, f_e)V_e}{n_{eff}}} \tag{2}$$

$$CI_{CE} = \sqrt{F_\alpha(1, f_e)V_e\left[\frac{1}{n_{eff}} + \frac{1}{R}\right]} \tag{3}$$

Where $F_\alpha(1, f_e)$ = The F-ratio at a confidence level of $(1 - \alpha)$ against DOF 1 and error DOF f_e

V_e = Error variance (from ANOVA)

$$n_{eff} = \frac{N}{1 + [\text{Total DOF associated in the estimate of the mean }]} \tag{4}$$

N = total number of results

R = sample size for confirmation experiment.

In equation 3, as R approaches infinity, i.e., the entire population, the value I/R approaches zero and $CI_{CE} = CI_{POP}$. As R approaches 1, the CI_{CE} becomes wider. These results are shown in Table 5.

CONFIRMATION EXPERIMENTS

The confirmation experiment is the final step in verifying the conclusions from the previous round of experimentation. The optimum conditions are set for the significant parameters (the insignificant parameters are set at economic levels) and a selected number of tests are run under constant specified conditions [14]. The results are shown in Table 5.

Table 4: Pooled ANOVA – raw data (out-of-roundness)

Source	SS	DOF	V	SS'	F-Ratio	F-Tab	P (%)	
A	0.0027	2	0.00135	0.0027	994.80	3.15	20.21	*
B	0.0030	2	0.00152	0.0030	1115.84	3.15	22.67	*
C	0.0056	2	0.00280	0.0056	2055.09	3.15	41.77	*
D	0.0002	2	0.00011	0.0002	81.45	3.15	1.64	*
E					Pooled			
A x B	0.0008	4	0.00021	0.0008	154.98	2.53	6.26	*
A x C	0.0003	4	0.00007	0.0003	52.39	2.53	2.09	*
B x C	0.0006	4	0.00015	0.0006	112.65	2.53	4.54	*
Error	0.0001	60	0.000001	0.0001			0.81	
Total	0.0134	80					100.00	

Symbols

SS = Sum of Squares, **DOF** = Degrees of Freedom, **V** = Variance,

SS' = Pure Sum of Squares, **P** = Percent Contribution, **A** = Work Piece, **B** = Tool,

C = Grit Size, **D** – Power Rating, **E** – Slurry Concentration and

* = Significant at 95% Confidence Level

Table 5: Predicted optimal values, confidence intervals and results of confirmation experiments

Performance Characteristics	Optimal Levels Process Parameters	Predicted Optimal Value	Confidence Interval (95%)	Actual Value (Average of 3 Conformation Experiments)
Conicity(CC)	A1, B3, C3, D1, E1	0.015	CI_{pop}:$0.014 < \mu_{CC} < 0.016$ CI_{CE}: $0.013 < \mu_{CC} < 0.017$	0.0153

Symbols

CI_{pop}: Confidence Interval for the Mean of the Population
CI_{CE}: Confidence Interval for the Mean of the Confirmation Experiment
A: Workpiece B: Tool C: Grit Size D: Power Rating E: Slurry Concentration

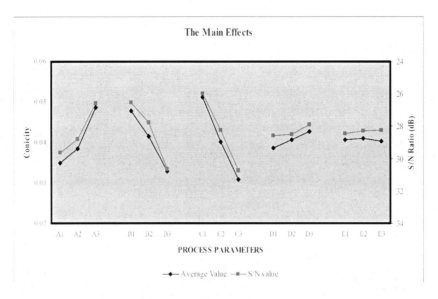

Figure 1 : Effects of process parameters on Conicity -raw data and S/N ratio: main effects
(A = Workpiece, B = Tool, C = Grit size, D – Power rating, E – Slurry concentration)

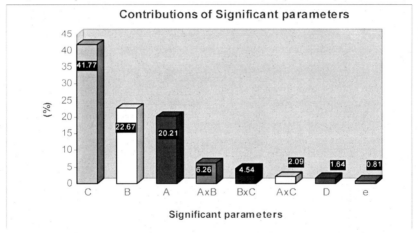

Figure 2 : Bar graphs showing percentage contributions of significant process parameters for
Conicity (CC): raw data
(A = Workpiece, B = Tool, C = Grit size, D – Power rating, E – Slurry concentration)

RESULTS AND DISCUSSIONS

The conicity increases with the increase in alumina content in the workpiece (Figure 1). This can be attributed to the fact that as the alumina content in the ceramics increases, the hardness and fracture toughness also increases which results into high tool wear. These results are in conformity with the studies of similar nature made earlier by different investigators.

In the present investigation, it has been observed that there is least lateral wear with TC tool and greatest with HCS tool. Also, the grooves were observed on tools used. These observations are in conformity with other investigators like, [5]. The tools can be ranked in the order of decreasing conicity as: HCS > HSS > TC (see Figure 2). Because of the high abrasion resistance, tungsten carbide tool has least wear as compared to high carbon steel and high speed steel. These results are similar to the findings of earlier investigators like [6, 9, 10].

From Figure 1, it can be seen that conicity increases almost linearly as the grain size increases from third level to second level and then to first level. TWR is the highest at first level and lowest for third level. This is due to the fact that the coarser grains cause more extensive damage of tool material during the impact of abrasives. This trend is similar to that of tool wear rate, which is the main cause of conicity in USM drilled holes.

The conicity increases marginally with the increase in power. This is due to the fact that power rating is associated with the amplitude of vibrations. As the amplitude of vibrations increases, the impact of abrasive with the tool also increases, resulting into higher wear rate.

The conicity increases marginally with increase in concentration from 25% to 30% to 35% but the increment is insignificant. This can be attributed to the fact that the tools used in the present investigation have high resistance to abrasion.

There is a significant interaction between tool & workpiece, workpiece & grit size and tool & grit size with regard to the raw data. However, the interaction between workpiece & grit size does not significantly affect the S/N ratio of conicity. Percentage contributions of different parameters are plotted and are shown in Figures 2 .

CONCLUSIONS

1. Taguchi's robust design method is suitable to analyze the ultrasonic drilling problem as described in this paper.
2. Conicity (CC) increases with the increase in alumina content in the work piece. The tools can be ranked in the order of decreasing conicity as: HCS > HSS > TC. The conicity increases almost linearly as the grain size increases. The conicity increases marginally with the increase in power and concentration. There is a significant interaction between tool & work piece, work piece & grit size and tool & grit size with regard to the raw data. However, the interaction between work piece & grit size does not significantly affect the S/N ratio of conicity. From ANOVA, it is clear that all the individual factors except slurry concentration have significant effect on conicity. The percentage contributions of parameters affecting both mean and variation in decreasing order are: grit size (36.51), tool (28.51), work piece (15.89), interaction B x C (8.88) and interaction A x B (06.86). The optimal levels of various process parameters for minimum conicity were:

   o Work piece material   50% Alumina
   o Tool   Tungsten carbide
   o Grit size   #500

- o Power rating   40%
- o Slurry concentration   25%
3. The predicted optimal range of conicity at 95% confidence level was 0.014 < CC < 0.016. The optimal results obtained were validated by conducting confirmation experiments.

REFERENCES

[1] Benedict, G. F. 1987. Nontraditional manufacturing processes. New York, Marcel Decker Inc. pp. 2-3, 67-86.

[2] Bellows, G. and Kohl, J. B. 1982. Drilling without drills. Am. Mach. Special Report No. 743. p 187.

[3a] Jadoun, R. S., Kumar, P., Mishra, B. K. and Mehta, R.C.S. (2006a) 'Optimization of process parameters for ultrasonic drilling (USD) of advanced engineering ceramics using Taguchi approach' *Engineering Optimization Journal*, UK, Vol.38, No. 7, pp.771–787.

[3b] Jadoun, R.S., Kumar, P., Mishra, B.K. and Mehta, R.C.S. (2006b) 'Manufacturing process optimization for tool wear rate in ultrasonic drilling (USD) of engineering ceramics using Taguchi's method' *International Journal of Machining and Machinability of Materials*, Vol. 1, No. 1, pp.94–114.

[4] Kennedy, D. C. and Grieve, R. J. 1975. Ultrasonic machining- A review. *The Production Engineer.* 54(9) : 481-486

[5] Adithan, M. and Venkatesh, V. C. 1976. Production accuracy of holes in ultrasonic drilling. *Wear.* 40(3) : 309-318.

[6] Smith, T. J. 1973. Parameter influence in ultrasonic machining. *Ultrasonics.* 11(5) : 196-198.

[7] Kremer, D. 1981. The state of the art of ultrasonic machining. *Annals of the CRIP.* 30(1) : 107-110

[8] Neppiras, E. A. 1956. Report on ultrasonic machining., *Metalworking Production.* 100 : 1283-1288, 1333-1336, 1420-1424, 1599-1604, .

[9] McGeough, J.A. 1988. Advanced methods of machining. New York, Chapman and Hall, ISBN 0412319705. pp. 170-198.

[10] Markov, A. I. 1966. Ultrasonic machining of intractable materials. London, Iliffe Books Ltd. 350 p.

[11] Rozenberg, L. D. et al. 1964. Ultrasonic cutting. New York, Consultants Bureau.

[12] Barash, M. and Watana, P. D. 1970. The effect of ambient pressure on the rate of material removal in ultrasonic machining. *Journal of machining science.* 12 : 775-779.

[13] Laroiya, S. C. et al. 1992. Machining of advanced ceramics. *In :* 15th AIMTDR Conference, December 3-5, 1992. pp. 571-575.

[14]Ross, Philip J. 1988. Taguchi technique for quality engineering. New York, McGraw-Hill book company

[15]Roy, Ranjit K. 1990. A primer on Taguchi method. New York, Van Nostrand Reinhold

[16]Byrne, D. M. and Taguchi, S. 1987. The Taguchi approach to parameter design. *Quality progress.* pp. 19-26.

SIMULATION OF MATERIAL REMOVAL RATE IN ULTRASONIC DRILLING PROCESS USING FINITE ELEMENT ANALYSIS AND TAGUCHI METHOD

R. S. Jadoun[1*], B. K. Mishra[2], Pradeep Kumar[2], R. C. S. Mehta[1]

[1]Production Engineering Department, College of Technology, Pantnagar-263 145 (INDIA)
[2]Mechanical and Industrial Engineering Department, IIT Roorkee, Roorkee-247667 (INDIA)

ABSTRACT
     Alumina being a lowest-cost high-performance ceramic is considered when seeking an alternate material for increased wear resistance, improved chemical resistance, dimensional stability, decreased friction and higher temperature use. Ultrasonic machining is a valuable process for the precision machining of such materials. In the present work, the mechanism of material removal by Silicon Carbide abrasives in alumina based ceramics with no sub-surface defects using USD process has been simulated. The Finite element method (FEM) software, ANSYS 5.4 is used for stress analysis under assumptions. The boundary conditions are applied. The theory of Zhang et al. (1999) for the number of abrasives under the tool is assumed to fit in the present model.
     The stress analysis confirms that the mechanism involved in material removal of brittle materials is initiation and propagation of the median vent cracks and lateral cracks. Fracture at the exit of hole was observed during experimentation. The material removal rate initially increases and then decreases with the increase of power input. The variation of MRR with respect to feed rate is very high, when compared to the variation due to power rating and slurry concentration. It is concluded that the Finite element model developed for the interaction of the abrasive and work piece is predicting the stress pattern with reasonable accuracy. The results of FEM analysis are validated by conducting experiments as per Taguchi's L_9 OA.

INTRODUCTION
     The twentieth century has witnessed the creation of products made from the most durable and consequently, the most un-machinable materials in the field of aerospace, nuclear science, automobiles and defense research. Inorganic materials are such materials being used in the above industries [2]. Out of these materials, advanced engineering ceramics have many attractive properties such as high hardness, high thermal resistance, chemical stability, and good insulating aspects. Alumina is the most mature high technology ceramic. Al_2O_3 can be used extensively for electrical applications due to its good electrical insulating characteristics. Because of the large quantity requirements, Al_2O_3 can be considered as the lowest-cost high-performance ceramic. Thus, they should be considered when seeking an alternate material for increased wear resistance, improved chemical resistance, dimensional stability, decreased friction and higher temperature use. However, the machining of Al_2O_3 by conventional manufacturing processes is extremely difficult and un-economical [3]. Ultrasonic drilling (USD) is a valuable process for the precision machining of hard, brittle materials such as Al_2O_3 because of many of it's unique characteristics.
     In spite of the fact that USD process has been used for many years, the mechanism of material removal is yet to be fully understood. All the theories propounded to date are either incomplete or controversial. [4]. Cook [5] and others [6-9] have done extensive work to

179

understand the mechanisms of material removal. On the basis of above findings, metal removal by means of USM is believed to be due to a combination of four mechanisms viz. Mechanical abrasion by direct hammering of the abrasive particles against the work-piece surface; Micro chipping by impact of the free moving abrasive particles; Cavitation erosion by abrasive slurry; Chemical action associated with the fluid employed [4,10].

The material removal takes place due to the individual or combined effect of above mechanisms by (i) shear, (ii) fracture, and (iii) plastic deformation [11-12]. Markov [13] and others [4,14] considered that first two mechanisms are primarily responsible for major stock removal while cavitation erosion and chemical effects were of secondary significance for normal materials machined by this process. However, in case of porous materials like graphite, cavitation erosion is a significant contributor to material removal.

Zhang et al.[15] analyzed the stress distribution during the ultrasonic drilling of holes in ceramics. The result show that in the terminal period of ultrasonic drilling, the stress on the periphery of the hole exit is at its maximum and is mainly responsible for fracture at the exit of holes in engineering ceramics. An effective method was also proposed to avoid such fracture. Wiercigroch et al.[16] postulated that the main mechanism of the enhancement of material removal rate (MRR) in ultrasonic machining is associated with high amplitude forces generated by impacts, which act on the work piece and help to develop micro-cracking in the cutting zone. The inherent non-linearity of the discontinuous impact process was modeled, to generate the pattern of the impact forces. A novel procedure for calculating the MRR was also proposed, which for the first time explains the observed fall in MRR at higher static forces. Rajurkar et al.[17] presented experimental simulation of the process mechanics in an attempt to analyze the material removal mechanism in machining of ceramic (Al_2O_3). It was found that low impact force causes only structural disintegration and particle dislocation. The high impact force contributed to cone cracks and subsequent crater change. Zhang et al. [1] analyzed the machining mechanism in the ultrasonic machining of engineering ceramics based on the fracture-mechanics concept. The study concluded that any increase in the amount of work / energy imparted to the ceramics in terms of the amplitude of the tool lip, the static load applied and the size of the abrasive would result in an increase in MRR. The results of study show that ultrasonic machining is an effective method for the machining of engineering ceramics. Ghahramani and Wang[18] discussed the USM process applications on the ceramic (Al_2O_3). Experimental simulations were used to analyze the mechanisms of material removal involving both the dynamic impacts of the abrasives from the high-frequency vibrations and the impact mechanism of abrasives forced by the hammering action of the vibrating tool. Jianxin and Taichiu[19] studied the effect of the properties and microstructure of the work-piece material on the MRR in ultrasonic machining of Alumina-based ceramic composites. Results showed that fracture toughness of the ceramic composite played an important role with respect to MRR.

Finite element method is a powerful tool in structural analysis of simple to complicated geometries, which uses discretization of a continuum into finite numbers of elements. Nodes of elements are assigned finite degrees of freedom of movement. Simple displacement functions are chosen for appropriate distribution of displacements over each element in conformity with compatibility at element boundaries. This powerful and analytical tool can account for different geometries of structures and different properties of each element. Banded stiffness matrix permits economical computations. In recent years with the coming of super computers the job of performing finite element analysis of a complicated geometry has become easy. ANSYS is one of such powerful tools in finite element method of analysis. Any complicated geometry can be analyzed easily using ANSYS.

Inspite of previous improvements achieved, there is a need to raise the scientific and technical knowledge for understanding the mechanism of the process, to improve its material removal rate. The effective process analysis and modeling help in understanding the influence of the parameters on the material removal rate. The computer aided finite element analysis can be considered to be one of the best solutions for this task. It is believed that, analysis of deformations and stresses generated at the workpiece and abrasive interface could provide deeper insight into the mechanism of material removal. Soundarya, et al. [20] proved experimentally that the direct impact of the abrasive particle on to the work surface causes the major material removal. So, in this present work direct impact of the abrasive on to the workpiece surface is modeled.

MODELING AND ANALYSIS PROCEDURES

FEM Formulation
It is impossible to exactly model the actual process. There is a need for assumptions depending upon the detail and accuracy required in modeling. The assumptions considered while modeling are:
1. The abrasive are of uniform size
2. The particles are cubic in shape
3. Only single layer of elements are distributed under the tool at any instant of time
4. The dominant mode of material removal is by the process of direct impact of the abrasive on to the workpiece
5. No loss of static load due to the resistance of the slurry
6. The materials properties are linear
7. The whole process is independent of temperature, i.e., temperature is constant
The Two models which are developed and analyzed for stresses are:

1. Model involving one particle under the direct impact between the tool and the workpiece, to study the effect of individual particles.
2. Model involving two particles under the direct impact between the tool and the work piece to study the effect of multiple impacts.
In general, the steps involved in both of the models are same and they are enlisted and explained in detail in the section 2.2. The ANSYS software which was selected for the FEM analysis is having the flexibility of using it either in batch mode of operation or in the graphical user interface (GUI) operation. Though both of them yield the same results, former one have the advantage of faster operation and used by advanced analysts. GUI will provide the ease of using the software for the beginners.

In this work a set of C programs were developed which generate the batch mode files for different model. These batch files are taken as input in the ANSYS and processed by it.

Steps in FEM modeling of this process
The steps involved in the modeling of this process are:
1. Planning the analysis
2. Determining the problem domain
3. Creating the finite element mesh

4.   Applying boundary conditions
5.   Applying the loads
6.   Convergence monitor
7.   Solving the problem
8.   Examining the results

The above steps are briefly discussed as under:

Planning the analysis
        The accuracy and cost (man-time, computer resources, etc.) of the analysis depends upon the planning decision. These two requirements are often in conflict. This is especially true for cutting edge analysis of very large models. Following have been considered in the planning phase of this modeling:

➢ Analysis type   :   Structural h-method
➢ Linear Vs Non linear   :   Material is assumed to exhibit linear elasticity
➢ Static Vs dynamic   :   Static

If the external excitation frequency of the force is less than the $1/3^{rd}$ of the structure's lowest natural frequency, the problem can be considered to be static. Modal analysis was carried out on the model and the structures lowest natural frequency is found to be 6802.4 Hz. Though, it is far less than that of the external excitation frequency, the problem is assumed to be static and analysis has been carried out.

➢ Objective of the analysis : To find the stress distribution in the work piece
                                 when the abrasive is compressed against the
                                 work piece.
➢ Units   : SI
➢ Element type   : Plane 42
➢ Material properties   : Alumina based ceramics

Determining the problem domain
        In this step the geometry of the problem is created. The geometry of the problem is a rectangular plate. The advantage of axi-symmetry can't be taken because 'n' numbers of particles are uniformly distributed under the tool and each take part in the total static load. The model is developed along the x-y axis. Considering, the tool to be vibrating about y axis. Only a small portion of the work piece i.e. 10-15 times the size of the abrasive is considered for analysis, since the effect of applied load will die out after this limit.

Creating the finite element mesh
        Since, in the finite element analysis, the results are available only on the nodes, the geometry of the problem is meshed such that the nodes can be found on the locations where the values of the degree of freedom (DOF) are needed. Also, at the regions, where the changes in gradient of the DOF are expected to be higher are meshed finer (at the contact). For fulfilling all these requirements mapped meshing is used. The relevant parameters for creating the finite element mesh are given below:

➢ Element type   :   Structural solid plane 42
➢ Dimension   :   2-D
➢ Shape   :   Quadrilateral, four nodes
➢ Degree of freedom   :   two at each node with translational in

> Capability : the x and y directions
Plasticity, creep, swelling, stress
stiffening, Large deformations and large
strain capabilities

Applying boundary conditions
The boundary conditions for the present problem are as follows:
1. At the base of the work piece
$U_x = 0$ and $U_y = 0$
2. The top of the abrasive the nodes are coupled i.e. displacement is same in the y
direction.

Applying the loads
Zhang et al. [1] have given a following formula for the number of abrasives under the
tool which is assumed to fit in the present model.

$$n = \frac{6\,SC}{\pi\,d^2} \qquad (1)$$

where,
S = Area of the bottom face of the tool
C = Concentration of the abrasive in the slurry
d = Mean diameter of the abrasive grain
Assuming these 'n' particles will simultaneously come into contact with the work piece. The
static load that was imparted to each abrasive is given by

$$F_y = \frac{Static\,Load}{n} \qquad (2)$$

$$P = \frac{F_y}{A} \qquad (3)$$

where,
A is the area of cross-section of the abrasive particle.
The convergence monitor for displacement is selected as 1.0 e-08 for accuracy in the
results. Number of global iteration values is set as 75. The solution is achieved whenever the
convergence monitor reaches at 1.0e-08.

Solving the problem
In the solution procedure, 'Frontal solver' is used. It involves:
a) After the individual element matrices are calculated, the solver reads the DOF for
the first element
b) The program eliminates any DOF that can be expressed in terms of the other DOF
by writing an equation to the file with extension TRI. This process repeats for all
elements until all degree of freedom have been eliminated and a complete
triangularized matrix is left on the file with extension TRI.
c) The frequently used term in frontal solver is 'Wave front'. The wave front is the
number of degrees of freedom retained by the solver while triangularization of the
matrix.

Conversion of FEM output into material removal rate

The results from any FEM analysis will be in the form of stresses and strains. These stresses are to be interpreted to determine the amount of material removed. A failure criterion should be selected which correlates the stresses with the possibility of failure, to predict which part of the material is failing due to the impact of the particle.

A number of theories of failure are described in the literatures which are used in design to predict the failure. As the mechanism involved in the material removal of brittle material is cleavage cracking. Modified Mohr's theory is used in the present investigation. A FORTRAN program is developed which takes the input from the FEM program and do the necessary analysis. This developed program gives results in the form of desired MRR. Since plane stresses are considered, the volume of material removed by each particle is given by the product of area and the size of the particle, 'v'. The tool will be vibrating at an ultrasonic frequency i.e. if 'f' is the frequency, tool will move towards the work piece f number of times. Thus there will be f number of impacts per second. Thus, total material removed is given by following equation

$$MRR = fvn \qquad\qquad (4)$$

where,

MRR	=	Material removal rate per second
f	=	Frequency of vibrations
v	=	Volume of material removed by each particle
n	=	Number of particles

RESULTS AND DISCUSSIONS

Finite element analysis involves three stages of activity: preprocessing, processing, and post processing. Preprocessing involves the collection of information, such as nodal coordinates, connectivity, boundary conditions, loading and material properties. Subsequent upon supplying the above information, processing is done, which involves solution of equations i.e. evaluation of elemental and nodal variables. The post-processing is concerned with the presentation of results. A complete finite element analysis is a logical interaction of these three stages. ANSYS software has been used for this purpose to obtain Nodal solution plots of the Von-Mises stresses; The principal stresses and shear stresses; Graphs of the stress variations with respect to the depth of the work piece; Variation of stresses at the depth of 50 µm for different loading and boundary conditions; Prediction of MRR from the developed FEM model. The results are discussed under Stress analysis and Prediction of MRR

Stress Analysis
Single impact model

Figure 1 shows the variation of the stresses developed in the subsurface due to the static force of a single abrasive particle hammered on to the surface of the work piece. Increasing stresses as the surface is approached from the subsurface are significant in initiating fractures and enhancing crack growth [18]. It is evident that the stresses will diminish out as we move from the point of contact. It is also evident that only material to certain depth is subjected to stresses which will exceed the fracture strength of the material and causes the material to be removed.

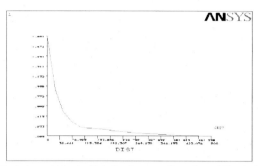

Figure 1: Stress Variation within the Sub-Surface as a Function of Depth from the Impact Zone

By comparing the present result with the previous microscopic observations of the surface damage from impact test some conclusions can be drawn. The relatively high compressive stress along the loading axis creates a plastic zone from a combination of micro structural coalescence and particle disintegration. Just below the plastic zone, median vent characteristic of cone cracks, develop as a result of the compressive stresses. This is consistent with the mechanism of cracking as proposed by [17, 19, 21, 22].

The tensile strength of alumina ceramics is nearly ten times less than its compressive strength. As a result, the tensile stresses developed at higher loads induce lateral cracks in the material subsurface. Lateral crack growth parallel to the surface of the material is observed to propagate under high force values [18]. The symmetry of the stress distribution is due to consideration of the single particle in the simulation. Maximum stresses for various loads are given below in the Table 1.

The Nodal stress distribution for the single impact model is obtained which shows the stress in the vicinity of impact causes the maximum stresses and material removal in the impact zone only.

Table 1: Results of Stresses Developed in the Sub-Surface for Single Impact (50 μm from surface)

S. No.	Load N	Max. Compressive Stress, MN/m^2	Max. Tensile Stress, MN/m^2	Max. Shear Stress, MN/m^2
1	0.02	205.59	8.37	61.02
2	0.03	308.38	12.56	91.55
3	0.04	411.17	16.74	122.04
4	0.05	521.43	20.93	152.56
5	0.06	616.76	25.12	183.07
6	0.07	719.55	29.31	213.58

3.1.2   Double Impact Model

The results from this model analysis could be extended to encompass the impact mechanism resulting from multiple impacts, as in the actual USM process. In the model the interaction effects resulting from such adjacent impacts were analyzed. This was carried out at the same static forces as applied in the single impact model. The variation of the stresses developed in the subsurface due to the static force of two abrasive particles hammered on to the surface of the work piece is obtained. There is some difference in planar stress distributions, between the single and double impacts in the tensile and shear stress distributions. The compressive stresses developed along the loading axis were high in both cases and were responsible for material removal by particle coalescence and disintegration in the plastic zone.

Material removal at the contact zone is mainly due to the impact force of the individual abrasive particles. Median cracks along the loading axis are likely to be present under each individual impact. However, the compressive stresses developed are independent of the adjacent impacts, and these impacts do not influence the characteristics of the damage. The Nodal stress distribution for the double impact model is also obtained. These stresses in the vicinity of impact cause the material removal in the impact zone only.

When compared to the single impact, the higher tensile stresses in the interaction zone between the adjacent impacts play a vital role in weakening of the material. This results in promoting lateral crack growth, and should enhance the material removal in the region between the impact locations. This should result in increased material removal in ultrasonic machining, but it is not being observed in actual machining process.

Interaction effect of multiple impacts

Considering a plane under the reference axis of loading, there is insignificant variation in the compressive stresses developed due to the double impact. Only the tensile and the shear stresses are affected by the influence of a double impact. Maximum stresses for various loads are given below in the Table 2.

Table 2 Results of Stresses Developed in the Sub-Surface for Double Impact
(50 μm from surface)

S. No.	Load N	Max. Compressive Stress, MN/m²	Max. Tensile Stress, MN/m²	Max. Shear Stress, MN/m²
1	0.02	215.34	11.71	70.39
2	0.03	298.42	17.58	105.92
3	0.04	427.69	23.43	140.18
4	0.05	540.77	27.32	173.99
5	0.06	595.82	35.16	210.34
6	0.07	734.54	41.03	248.13

The tensile stresses in the region show a substantial increase of around 25 to 30%. But Ghahramani et al. [18] reported a decrease in shear stresses in the multiple impacts, which may be due to decrease in material removed by each individual impact. The variation in the results may be due to linearity in the present model.

In the double impact model, the shear stresses in the interaction zone were lower than the stresses developed from the single impact, which may be attributed to the multiple directional characteristics from each individual impact. This may be one of the reasons for the lower material removal exhibited by ultrasonic machining.

Effect of the distance between impacts

The stresses at any specific node in the finite element model are observed to be increasing proportionally as the value of static load is increased. When the stresses developed at specific locations were compared for the single and double impacts, the effect of an adjacent double impact was minimal.

Taking the node under the right impacting particle as the reference node, the value of Von-Mises stresses as calculated are given in Table 3. The results show that the Von-Mises stresses are independent of the type of impacts.

Table 3: Von-Mises Stresses under Single and Double Impacts
(45µm abrasive, 30 N Static Load)

S.No.	Type of Impact	Von-Mises Stresses (MPa)
1	Single	183.7.5
2	Double	183.502

The Von-Mises stresses in the interaction zone are at their maximum when the loading axis is at an optimal distance from each other, and as the impact locations progressively became closer, these stresses in the zone between the impacts become lower than the stresses in the region outside.

These results also confirm the low material removal rates observed when the ultrasonic machining process operates under very high or extremely low slurry concentrations. It is, therefore, possible to conclude from these results that at an optimal distance between the impacting abrasives, the maximum compressive, and tensile stresses are developed. However, if the indentation locations of the impacting particles are very close or extremely far apart, the stresses developed are reduced to a minimum. Hence, little or no material removal will occur from multiple impacts during ultrasonic machining process when the impact locations are either very close or extremely far apart [18].

Effect of abrasive grain size

In the present investigation, an attempt has been made to simulate the stresses developed in the sub-surface by increasing the size of the abrasive grains. The stresses are symmetrical about the loading axis. The stresses at any specific plane in the FEM model are observed to be increasing almost linearly as the size of the abrasive is increased. With the increase of abrasive size, the number of impacting particles decreases and thus the static load per particle increases, resulting into the increase in the stresses as observed from the model.

Prediction of material removal rate from FEM

It can be observed from the results obtained that the material removal rate depends upon the following:
1.	Size of the abrasive
2.	Slurry concentration
3.	Static load
4.	Distance between the abrasives

Effect of the size of the abrasive

The material removal rate is found to be increasing as the size of the abrasive is increased. The reason for this increase of MRR is the dependence of the model on the maximum compressive stresses that were developed in the interaction zone. The maximum compressive stresses increase as the size of the abrasive increased. The material removal rate decreases when the grain size is sufficiently large. Thus, there is a limit to the effect of grain size on the rate, as a very coarse powder cause a reduction in the MRR. The effect of the size of the abrasive on MRR predicted by this model is in tune with the similar studies using different analysis tools as reported in the literature [1, 12]. The failure criterion adopted for this model predicts crack under gradual loading conditions. Whereas, in the actual machining process, because of the quick loading, there is no sufficient time for the crack to propagate. Secondly, static analysis has been done. Thus, the MRR predicted by the model is comparatively high. It was reported that the 99% Al_2O_3 will have a machining rate between 5.4 and 9.0. This model predicts 2-3 times higher MRR due to above explained reasons.

Effect of concentration

The material removal rate predicted by this model is found to be increasing linearly as the concentration of the slurry increases. However, it was reported in the literature that MRR should increase up to optimum point and then decreases for further increase in the concentration. This discrepancy may be due to assumed linearity of the model.

Validation of model through experimentation

The validation of the FEM model is carried out through experimentation as per the Taguchi's method of robust design. The robust design methodology, proposed by Taguchi [23, 24] is one of the appropriate methods for achieving this goal. The purpose of the robust design methodology is to determine optimal settings of design or controllable variables in a system such that the system's performance is robust to various noises. The experiments are carried out using 50% alumina based ceramic composite, silicon carbide of the 320 grit size and high carbon steel tool.

The three process parameters viz. power input; feed rate and slurry concentration have been selected. Experiments are conducted as per the test conditions specified by the L_9 OA. Each experiment was repeated thrice in each of the trial conditions. The flow rate of the abrasive slurry was maintained constant for all the experiments. Thus, twenty seven holes are drilled by USM process.

Analysis of experimental results and discussion

The experimental results are analyzed and analysis of variance (ANOVA) is performed. The analysis of raw data variance with arithmetic average of material removal rate was made with the objective of analyzing the effects of amplitude, feed rate and concentration on the total variance of the results. The analysis has been carried out at confidence level of 95%.

From the ANOVA of MRR raw data, it is clear that

(i) The feed rate significantly affects the mean value of MRR.

(ii) The power input and concentration are insignificant.

So the consideration should be given while selecting the parameter feed rate. It should always be selected at the optimum level. The rest of the parameters can be set at any economic level. This is in confirmation with the results of Zhang et al.[1] in which they concluded that the feed rate / static load is causing the maximum variation in MRR. The ANOVA of S/N ratio given in Table 5 is used to select the optimum value of the parameters.

CONCLUSIONS

The aim of the present work was to develop a finite element model for the Ultrasonic machining at the abrasive workpiece interaction zone and to analyze the stress pattern to predict the material removal mechanism.

Based upon the Investigations carried out following conclusions can be drawn

1.   The study of mechanism of material removal in alumina based ceramics with no subsurface defects is carried out using Finite element software, ANSYS. Here, ANSYS is used to model the problem and to evaluate the stresses. The nodal solution plots of the Von-Mises stresses, the principal stresses, and the shear stresses are obtained. Graphs of the stress variation with respect to the depth of the workpiece are also obtained. The material removal rate is predicted from the developed FEM model using FORTRAN program. The stress analysis confirms that the mechanism involved in material removal of brittle materials is initiation and propagation of the median vent cracks and lateral cracks. Fracture at the exit of hole was observed during experimentation.

2.   The results of Finite element model have been validated by conducting phase-III experimentation. The process parameters viz. power rating; feed rate and slurry concentration are selected and varied as per the test conditions specified by Taguchi's L_9 OA. The material removal rate initially increases and then decreases with the increase of power input. The variation of MRR with respect to feed rate is very high, when compared to the variation due to power rating and slurry concentration. It is concluded that the Finite element model developed for the interaction of the abrasive and work piece is predicting the stress pattern with reasonable accuracy.

3.   The developed double impact model reveals that the reduced shear stress when compared to single impact may be one of the reasons for the reduction in MRR.

REFERENCES

[1]   Zhang, Q. H. et al., "Material-removal-rate analysis in the ultrasonic machining of engineering ceramics", *Elsevier, Journal of Materials Processing Technology,* **88,** 180-184 (1999).

[2]   Benedict, G. F, "Nontraditional manufacturing processes", New York, Marcel Decker Inc., pp. 2-3, 67-86 (1987).

[3]   Pandey, P. C. and Shan, H. S., "Modern machining processes", New Delhi, TMH Publishing Co. Ltd., pp. 2-38 (1980).

[4]   Shaw, M. C., "Ultrasonic grinding', *Microtechnic,* **10(6),** 257-265 (1956).

[5]   Cook, N. H., "Manufacturing analysis", New York, Addison-Wesley. pp. 133-138 (1966).

[6]   Neppiras, E. A. and Foskett, R. D., "Ultrasonic machining – II: Operating conditions and performance of ultrasonic drills', *Philips Technical Review,* **18(2),** 368-379 (1957).

[7]   Graff, K. F., "Macrosonics in industry: ultrasonic machining", *Ultrasonics,* **13,** 103-109 (1975).

[8]   Kainth, G. S. et al., 'On the mechanics of material removal in ultrasonic machining", *International Journal, MTDR,* Pergamon Press, **19,** 33-41 (1979).

[9]   Kremer, D., "The state of the art of ultrasonic machining", *Annals of the CRIP,* **30(1),** 107-110 (1981).

[10] Weller, E. J., "Non-traditional machining processes", *Society of Manufacturing Engineers,* **2,** 15-71 (1984**).**

[11] Khairy, A. B. E., "Assessment of some dynamic parameters for the ultrasonic machining process" *Wear,* **137,** 187-198 (1990).

[12] Pentland, E. W. and Ektermanis, J. A., "Improving ultrasonic machining rates – some feasibility studies" *Journal of Engineering for Industries,* Transactions of the ASME, **87(series B),** 39-46 (1965).

[13] Markov, A. I., "Kinematics of the dimentional ultrasonic machining method", *Machines and Tooling,* **30(10),** 28-31 (1959).

[14] Neppiras, E. A., "Report on ultrasonic machining", *Metalworking Production,* **100,** 1283-1288, 1333-1336, 1420-1424, 1599-1604, (1956).

[15] Zhang, Q. H. et al., "Fracture at the exit of the hole during the ultrasonic drilling of engineering ceramics", *Elsevier, Journal of Materials Processing Technology,* **84,** 20-24 (1998).

[16] Wiercigroch, M. et al., 'Material removal rate for ultrasonic drilling of hard materials using an impact oscillator approach", *Elsevier, Physics Letters,* **A 259,** 91-96 (1999).

[17] Rajurkar, K. P. et al., "Micro removal of ceramic material (Al$_2$O$_3$) in the precision ultrasonic machining", *Elsevier, Precision Engineering,* 23 , 73-78 (1999).

[18] Ghahramani, B. and Wang, Z. Y., "Precision ultrasonic machining process: a case study of stress analysis of ceramic (Al$_2$O$_3$)", *Pergamon, International Journal of Machine Tools and Manufacture, 41,* 1189-1208 (2001).

[19] Jianxin, D. and Taichiu, L., "Ultrasonic machining of alumina-based ceramic composites", *Journal of European Ceramic Society,* **22,** 1235-1241(2002).

[20] Soundarajan, V. and Radhakrishnan, V., "An experimental investigation on the basic mechanisms involved in ultrasonic machining", *International Journal, MTDR,* **26(3),** 307-321, (1986)

[21] Evans, A. G. and Wilshaw, T. R., "Quasi-static solid particle damage in brittle solids-I: observations, analysis and implications", *Acta Metalurgica,* **24,** 939-956 (1976).

[22] Komaraiah, M. and Narasimha Reddy, P., "A study on the influence of work-piece properties in ultrasonic machining", *International Journal of Machine Tools Manufacturing,* **33(3),** 495-505 (1993).

[23] Taguchi, G., "Quality engineering in production systems", New York, Mcgraw-Hill publishing House (1989).

[24] Phadke, M. S., "Quality engineering using robust design", New Jersey, Prentice Hall (1989).

Author Index

Author Index